THE YOUNG ENGINEER BOOK OF SUPERCARS

The Japanese Datsun 260Z is the world's top-selling sports car. It comes in both two and four seat versions. The four-seater 'Z', shown here, has a 6-cylinder engine that propels it at up to 200 kph, with a 0-100 kph time of about 9 seconds. The car is tough too—modified versions have twice won the gruelling East African Safari Rally in appalling conditions.

Mercedes-Benz of Germany built this C-111 research car in 1969 to test the then-new Wankel rotary engine. Although this type of engine has not yet proved very successful, the C-111 nevertheless could reach 260 kph. Its unusual 'gull-wing' doors were an echo from an earlier design—they were also used on the 1952 Mercedes-Benz 300 SL.

The Italian Ferrari 512 BB Boxer is powered by a 12 cylinder engine just in front of the rear wheels. The 1,515 kg car can travel at around 290 kph. The performance is heavy on fuel though—a litre of petrol only keeps the Boxer going for a little over 5 km.

Written by
Jonathan Rutland
Art and editorial direction
David Jefferis
Text editor
Eliot Humberstone
Design
Iain Ashman
Technical consultant
Roger Ames
Illustrators
Malcolm English
Terry Hadler
John Hutchinson
Frank Kennard
Jack Pelling
Michael Roffe
Special photography by
Peter Mackertich

Acknowledgements
We wish to thank the following individuals and organizations for their assistance.

Airfix Products Ltd
Brands Hatch Racing Ltd
Children's Research Unit
Chrysler UK Ltd
Institute of Advanced Motorists
Tony Lanfranchi
Lotus Cars Ltd
Maranello Concessionaires Ltd
Lauris Morgan-Griffiths
Panther West Winds
Porsche Cars GB Ltd
Project Thrust
Nancy Rew-Smith
Royal Automobile Club
Santa Pod Raceway
Glen Smith
Under 17s Car Club
Barrie Williams

First published in 1978 by
Usborne Publishing Ltd
20 Garrick Street
London WC2E 9BJ

Published in Australia by
Rigby Ltd
Adelaide, Sydney,
Melbourne, Brisbane,
Perth

Published in Canada by
Hayes Publishing Ltd
Burlington, Ontario

Printed in Belgium by
Henri Proost, Turnhout,
Belgium.

INTRODUCING SUPERCARS

These pictures give you an idea of the parts which go to make up a modern supercar, in this case a Lotus Elite.

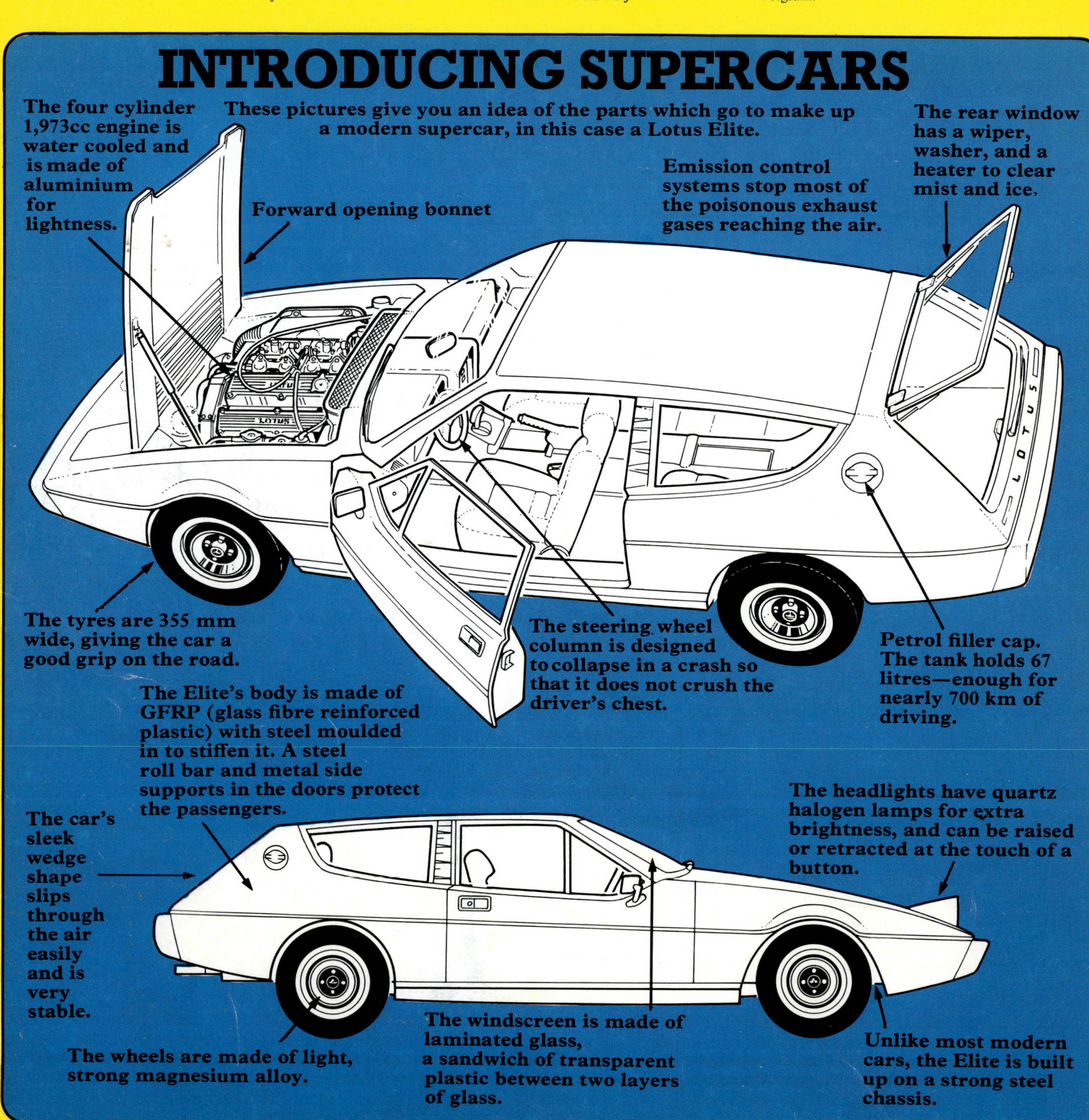

THE YOUNG ENGINEER BOOK OF SUPERCARS

ABOUT THIS BOOK

Supercars are the fastest, surest handling, quickest accelerating, best looking thoroughbreds of the motoring world. They are the most impressive machines on the road or at the race track.

Throughout the history of the motorcar, there have been cars that have stood out as something special. It may have been because of their exotic bodywork or superlative performance or, in some rare cars, an ideal blend of both. Supercars may be built by large international companies, or by small specialist firms making only luxury models. Sometimes they are produced as 'one-off' show cars to test new designs.

This book goes inside the supercar, explaining how the various components go together, how the cars have developed over the years, and shows some of the most fantastic design ideas that have been tried out.

Simple experiments show you some of the principles behind car engineering, and you can even try your hand at designing your own supercar.

Mel Nichols
Editor, *CAR* magazine

CONTENTS

EARLY MOTORCARS

The earliest ancestor of the motorcar was built by a French engineer, Nicholas Cugnot. His vehicle, shown in the first picture on the right, was steam powered and not very successful. Other steam vehicles were tried during the 19th century, but the slow and cumbersome machines never achieved widespread use.

The breakthrough came in 1885, when a German engineer, Karl Benz, made the first car to be powered by a petrol engine. This was the start of the motoring revolution—from this moment, slowly at first, the world got used to being mobile.

1

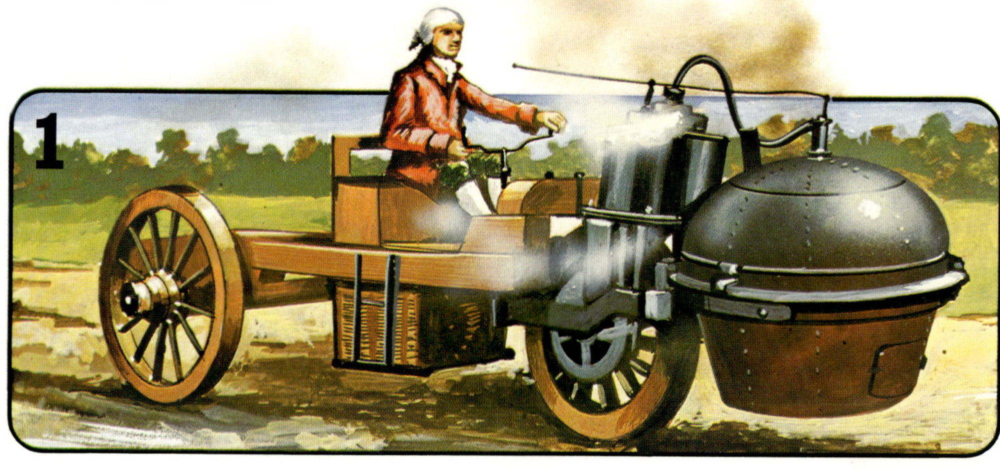

▲ Cugnot's steam carriage of 1769 had a front mounted tank to boil water in. The steam powered the car through a system of piston, cylinder, rods and cranks. It was designed to haul cannons for the French army, but its usefulness was limited by its performance. Its top speed was 5 kph and every 15 minutes it had to stop to build up steam again. The heavy boiler made it hard to steer and the first carriage crashed. He built another, but the army had lost interest.

4

Front engine, rear wheel drive

Mid engine, rear wheel drive

Rear engine, rear wheel drive

Front engine, front wheel drive

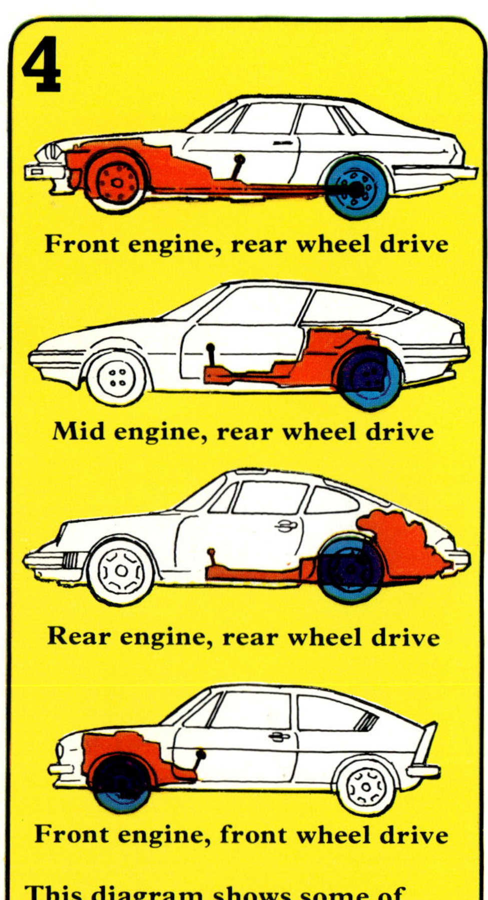

This diagram shows some of the different ways in which the engine and driving wheels can be arranged. From top to bottom the cars are: Jaguar XJ-S, Matra Bagheera, Porsche 911, Alfasud. Front engine, rear wheel drive has been the most popular, but front engine, front wheel drive is becoming standard for new cars.

5

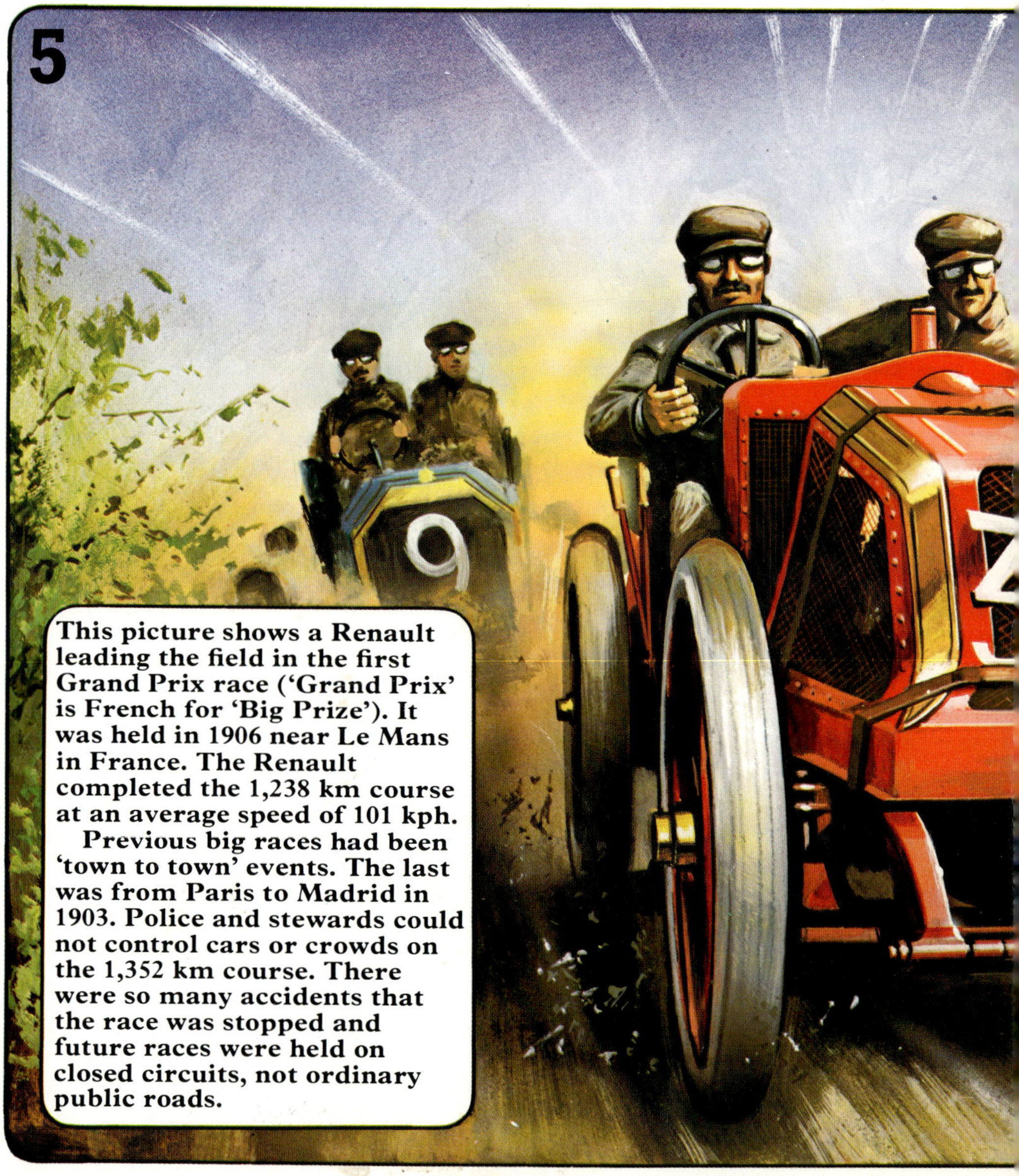

This picture shows a Renault leading the field in the first Grand Prix race ('Grand Prix' is French for 'Big Prize'). It was held in 1906 near Le Mans in France. The Renault completed the 1,238 km course at an average speed of 101 kph.

Previous big races had been 'town to town' events. The last was from Paris to Madrid in 1903. Police and stewards could not control cars or crowds on the 1,352 km course. There were so many accidents that the race was stopped and future races were held on closed circuits, not ordinary public roads.

In 1902 the French government tried to make motorists use alcohol fuel made from potatoes. The scheme was not a success.

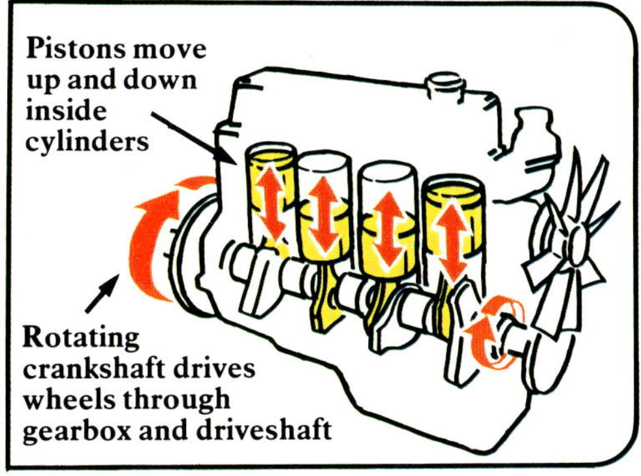

▲ This picture shows the first practical petrol-engined car, built by Karl Benz in 1885. On the right is a view inside a typical modern engine, based on Benz's principles. Petrol vapour is ignited at the top of each cylinder by an electric sparking plug. The explosions force the pistons down the cylinders. The vertical movements are changed to a rotary one by a rod and crankshaft system. The pistons move in sequence, providing a smooth flow of power.

▲ This 1891 Panhard Levassor pioneered the layout used in most cars ever since. The engine was at the front under a bonnet. The car had a gearbox, foot-operated clutch, rear wheel drive and a radiator to cool the water.

▲ The Ford Model T was the first car to be made on an assembly line. Before 1908 all cars had been hand made and very expensive. Mass production techniques enabled the time—and the cost—of car making to be drastically reduced.

▲ As the car's popularity grew, so did the number of makes. Soon there were over 500 in the USA alone. The three whose badges are shown above all became part of the American General Motors, now the biggest car maker in the world.

▲ This supercar of 1911 was the ancestor of all high performance sports cars. Built for the racing driver René de Knyff, it was called the Skiff, because of its boat-like appearance. Panhard Levassor made the basic car. The Skiff's revolutionary feature was its body, which made a complete break from the solid and heavy coachwork of all earlier supercars. It had the lightweight construction of aeroplanes of the time. The spoked wheels were used to reduce weight.

Henry Ford's first car of 1896, was the only car in Detroit. He had to chain it to lampposts when he left it unattended.

UNDERSTANDING SUPERCARS

The key to understanding how a car works is that it is made up of 7 systems, described in the small red boxes.

The large picture shows a skeleton view of a 1930 Bentley racing car. You can compare some of the differences—and similarities—between it and the 1970s car in the boxes.

Engine

The heart of the car. Engine controlled by floor-mounted accelerator pedal.

Transmission

Links engine to drive wheels by clutch, gearbox, propeller shaft, differential.

This handbrake, typical of the period, was mounted on the outside of the car. All cars now have handbrakes inside, usually between the front seats.

A side-view of the 217 kph Bentley

Propeller shaft

Front wheels mounted on a single axle, linked to the chassis by springs. Today all cars have independently sprung front wheels to improve ride.

Early cars had oil or gas lamps which gave a poor light output. Electric ones like this date from around 1913.

The capacity of the engine is measured in cubic centimetres (cc) or in litres. The size, in this case 4.5 litres, refers to the total volume of all the cylinders.

Early cars had a wood or metal structure called the chassis onto which the body was mounted. Few cars now have one— the body is built in one unit.

The tyres are air-filled (pneumatic). The first cars had solid tyres which gave a hard bouncy ride and poor grip, which reduced cornering and braking power.

Steering

Rod and lever system to turn the front wheels left or right. Controlled by steering wheel.

Brakes

Floor-mounted footbrake slows all wheels. Hand brake used for parking.

Electrics

Large battery provides power to start engine, work equipment such as lights, wipers and so on.

A saloon car is made of nearly 60 materials from steel and nickel to plastic and leather.

Body

In almost all present-day cars made of pressed steel units, welded together.

The large rear-mounted fuel tank carries petrol, fed by an electric pump along pipes to the engine. Modern tanks are placed in less exposed positions to reduce fire risk in the event of an accident.

The differential system is joined to the engine by a propeller shaft. The way of transmitting variable power to the two driving wheels remains the same today.

Suspension

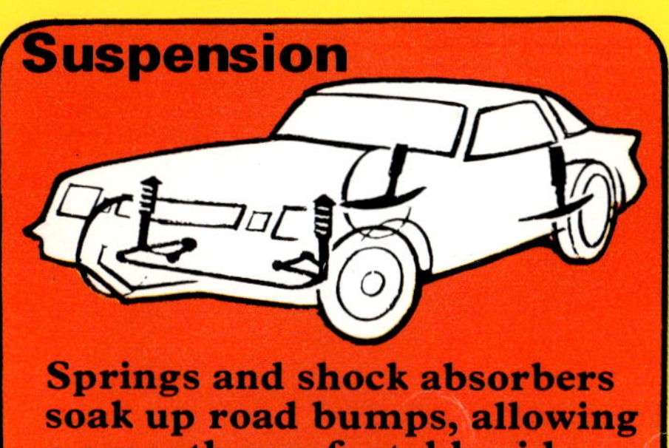

Springs and shock absorbers soak up road bumps, allowing a smooth comfortable ride.

Why a differential is used

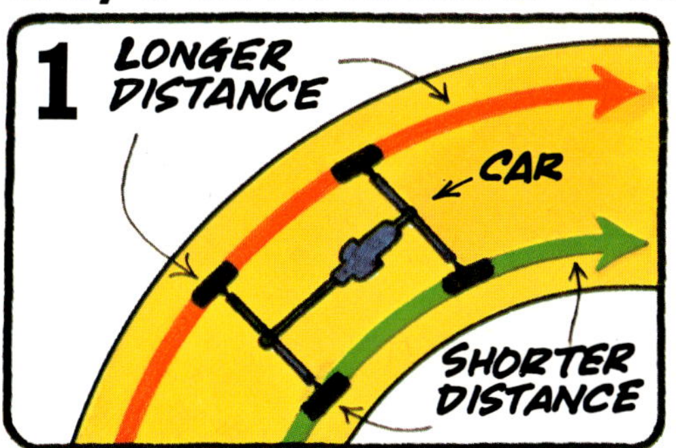

▲ When a car goes round a corner, the distance travelled by the wheels on either side varies. The outer wheels travel further than the inner ones. If the drive wheels were joined by a solid axle, one or both of them would slip and skid.

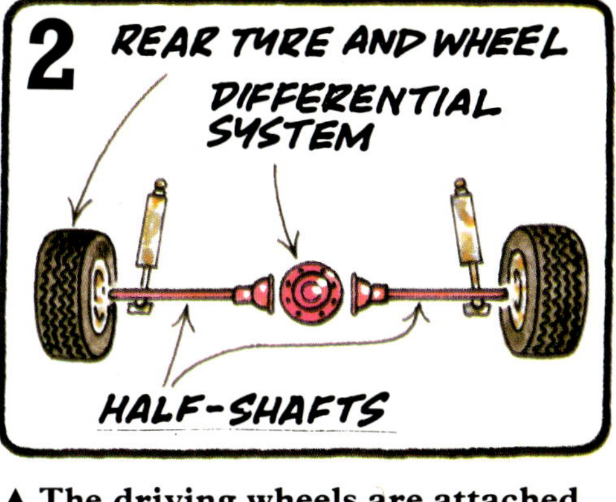

▲ The driving wheels are attached to an axle split into two halves. In the middle is a gear system called the differential. This allows the wheels to run at different speeds, allowing for the different distances they travel round bends.

▲ You can demonstrate the principles involved with this simple experiment. You need two pencils or (as we used) ball point pens, a roll of clear sticky tape and two cotton reels. If you cannot get these, try two plastic bottle tops.

▲ First you need to make a 'bandage' with some tape. Stick two pieces together as shown, so that the ball point pen can slide freely inside. Next, force the ends of the two pens into the holes in the cotton reels.

▲ Slide the two ball-pen half-shafts into the tape bandage. Mark a large dot on each pen, then cover each dot with a piece of tape. This will protect the dots and stop the centre 'differential' bandage from slipping along one pen.

▲ Rotate the wheel and axle assembly round in a tight corner, counting the number of turns each dot makes. You should find that the outer axle is rotating about four times as much as the inner one.

In six years an average car travels nearly 97,000 km. Each wheel will have turned 95 million times.

CLASSIC CARS

Motoring began as an uncomfortable, noisy and often messy adventure. But before long engineers began designing cars 'for the day after tomorrow', the slogan of the first Mercedes. In 1906 Rolls Royce brought out the Silver Ghost, the first really smooth, quiet and reliable car. Other makers followed their example, and on these pages we give our choice of the best cars of all time. All are famous for their comfort, style, performance and reliability. There have been many other supercars, each with their own combination of qualities. Try making your own list. You will find you have a vast selection of cars to choose from.

▲ Duesenberg SJ, 1932. Supercharged 6.9 litre engine, top speed 208 kph.

The supercar of American film stars and European royal families. By 1937, when the Duesenberg company collapsed, just 470 SJs had been made.

▲ Rolls Royce Silver Ghost, 1906. 7 litre engine, top speed around 125 kph.

To show its superior quality, the first Silver Ghost was driven day and night for 24,120 km. It had to make only one repair stop, and that took just one minute. The Ghost above actually had silver-plated fittings.

▲ Mercedes-Benz Model S, 1927. 6.8 litre engine, top speed 160 kph.

The Model S sports car was faster than Mercedes' GP racing cars of the time, and proved almost unbeatable on the track. 'S' cars were entered in races as late as 1938. Its good brakes, steering and roadholding enabled it to reach 160 kph in comparative safety.

► Jaguar XK 120, 1948. 3.4 litre engine, top speed 212 kph.

The XK 120 was the first of a new generation of streamlined high performance sports cars. It won race after race, yet was very popular as a road car and fairly cheap to buy. Its engine design is still in use today.

To pass different countries' various regulations, a Rolls-Royce has to pass 204 separate tests.

BUGATTI

▲ Bugatti Royale, 1927. 12.7 litre engine, top speed 193 kph.

The longest car ever made, built for kings and guaranteed for life. It was too large and too expensive—only six were made, three sold. The Royale had just three gears: one for moving off, one for normal driving and an overdrive for very high speeds.

▼ Chevrolet Corvette, 1953. Top speed 216 kph plus.

The Corvette is the USA's only true sports car. It is now in its third body style. The current one dates from the Mako Shark show car of 1967 shown below. Chevrolet has been making cars since 1912.

CHEVROLET

Classic cars revived

The car above does not look as if it was made in 1978, but the old-fashioned looking Panther Lima is as new as the sleekest Lamborghini.

Cars with the styling of the 1920s and '30s are a money-spinning part of the car-making business and Panther make several models, including one that looks like the Bugatti Royale shown on the left.

Interesting mixture

The Lima is a mixture of careful craftsmanship and parts bought 'off the shelf' from Vauxhall, the British subsidiary of American General Motors. Vauxhall parts include the 2.3 litre engine and instruments. The paintwork of our test car was yellow and black. Superbly applied, it was finished off by hand-painted stripes.

The car is a two-seater and has no proper boot—just a hole behind the rear seats under the rear-mounted spare wheel.

Squeezing in

The Lima is not an easy car to get into, at least with the hood up. The technique is: open the door, ease into the seat, curl up your legs and squirm round to face the front. Once there, it is very comfortable, with lots of leg room.

On the road

Performance is good, though the suspension is hard and the seat position low. Your head is little more than a metre or so above the ground, so 50 kph feels like 100. Even so, the big engine and light weight mean that the Lima will outrun most other cars.

Conclusion

The Lima is useless if you carry lots of baggage, but as a fun car it is super.

Rolls-Royce seats are tested too—a machine called IRMA 'squirms' a million times to check that seats will not collapse.

SUPERCAR DRIVING

What is it like to drive round a Grand Prix track in a racing car? To find out, we visited the Brands Hatch International circuit to be driven by top instructor Barrie Williams.

Motor racing is dangerous, and the driving methods shown here are for the track only, not for public roads.

Also, speed alone will rarely win a race. Skill, responsibility and safe driving are required in equal measure.

The diagram below shows the circuit, together with the line taken to get round it fastest, the clipping points of each bend (see frame 7) and the gears a driver will use.

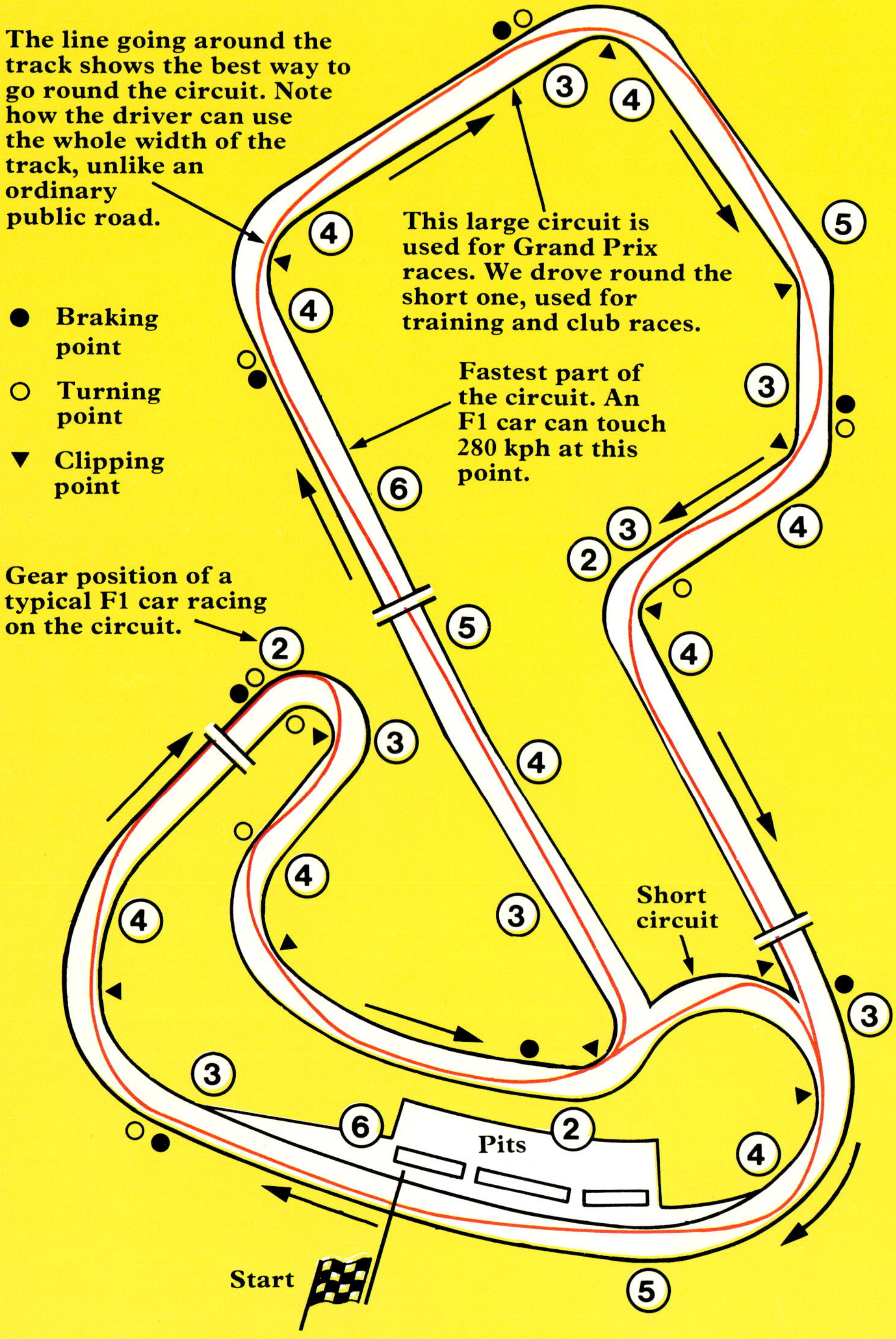

The line going around the track shows the best way to go round the circuit. Note how the driver can use the whole width of the track, unlike an ordinary public road.

This large circuit is used for Grand Prix races. We drove round the short one, used for training and club races.

● Braking point

○ Turning point

▼ Clipping point

Fastest part of the circuit. An F1 car can touch 280 kph at this point.

Gear position of a typical F1 car racing on the circuit.

Short circuit

Pits

Start

▲ Brands Hatch Racing runs several types of cars for training would-be drivers. These include single-seater Elden Formula Fords, two-seater Lola sports racers like the one shown in the picture above, and Ford Escort saloon cars.

▲ Here is a racer's eye-view of the track as Barrie Williams overtakes a pair of slower cars. Safety rules are strict for practise—overtaking is only allowed on the left. In a proper race, drivers get past slower competitors as best they can.

▲ Tyres screaming, the Escort rounds a bend, accelerating out of it down the straight. If the tyres get too hot under the strain of fast cornering they lose their grip. An experienced driver eases off a little to allow them to cool down.

The longest marathon run was in 1935. French driver Francois Leçot covered 400,000 km in daily runs, 19 hours a day.

▲ Full harness seat belts are essential in a racing car. The four-point system keeps a driver firmly in the seat. This picture shows a driver being strapped into a Formula-Ford. Crash helmets are always worn.

▲ The Ford Escorts used for race training are very different to the ordinary saloon. Modifications include a strong protective roll-cage, tuned engine and suspension system, alloy wheels and specially made racing tyres.

▲ The large red switch in this picture is for cutting out the car's electrical system instantly, to reduce the fire risk in an accident. If the driver is unconscious, a similar knob on the car's bonnet can be pulled by the rescue team.

7 Points of a bend

In the diagram below you can see the three points that a driver must learn for each corner on a circuit. Exactly when to brake, where to start turning and which part of the bend to aim for is knowledge a good driver needs to keep up the maximum safe speed.

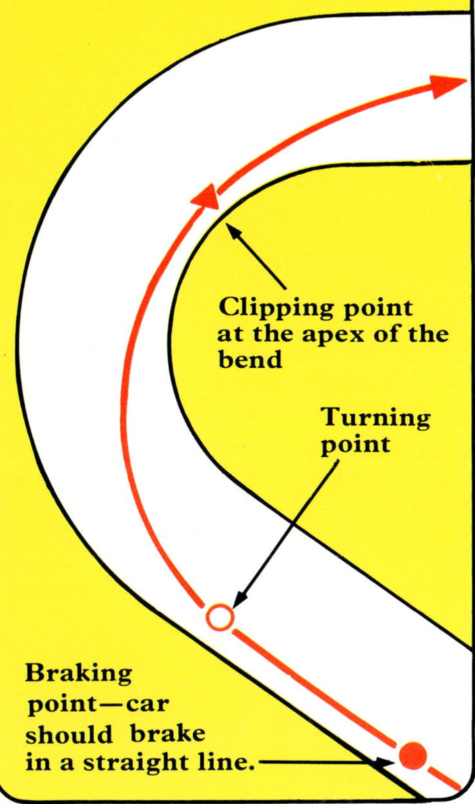

8 Sliding

As the car passes the turning point, the driver flicks the steering into the bend, pressing hard on the accelerator. This spins the back wheels, bringing the back of the car round. Then he turns the steering wheel the other way to keep the car on course.

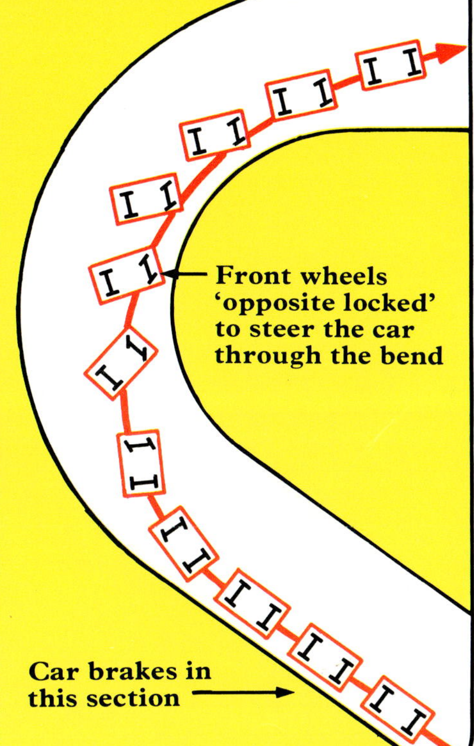

9 Drifting

In a drift, all four wheels are sliding at an angle to the direction of travel. The front wheels are turned sharply into the bend. A well-controlled drift will result in the car passing the clipping point at the correct angle to accelerate down the next straight.

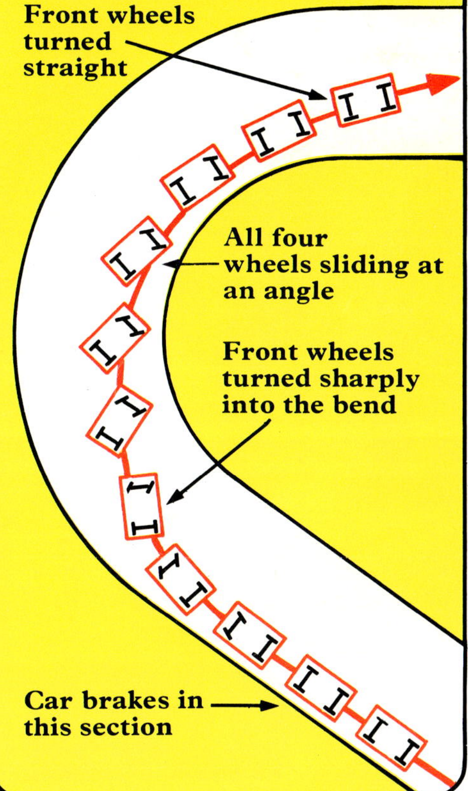

The Ford-Cosworth engine is the most successful racing engine ever made. First designed in 1966, it still powers most F1 cars.

Oil additives

Tyres

BRUT

Men's cosmetics

Marlboro

Cigarettes

GRAND PRIX RACING

Every race has a formula. This specifies the maximum size and weight of cars, engines, safety equipment, and even the length of the races themselves. The kings of the racetrack are the Grand Prix (GP) Formula 1 single seaters with engines of up to 3,000 cc. Most Formula 1 cars use the same engine, a Cosworth-Ford. Developing, building and racing cars is enormously expensive, so sponsors are needed to help pay for the cars. You can see some of their names around this page.

► On the right are some of the best GP cars. The 1912 Peugeot (1) had a small engine, but raced successfully against 14 litre Fiats and other giants. The 1937 Mercedes-Benz W125 (2) was the most powerful GP car ever. The Maserati 250F (3) was a classic GP car of the 1950s, while at the bottom (4) is a side view of the 1976 world champion Ferrari.

PIRELLI

Tyres

Front and rear aerofoils work like upside-down aircraft wings. At speed, airflow creates downthrust keeping the car firmly on the road. A rear aerofoil like this creates about 180 kg of downthrust.

Ferrari is one of the most famous and successful names in GP racing, and one of the few to design both car and engine. This is their 1976 world champion car.

The cockpit is so small, the driver has to squeeze in with his arms over his head.

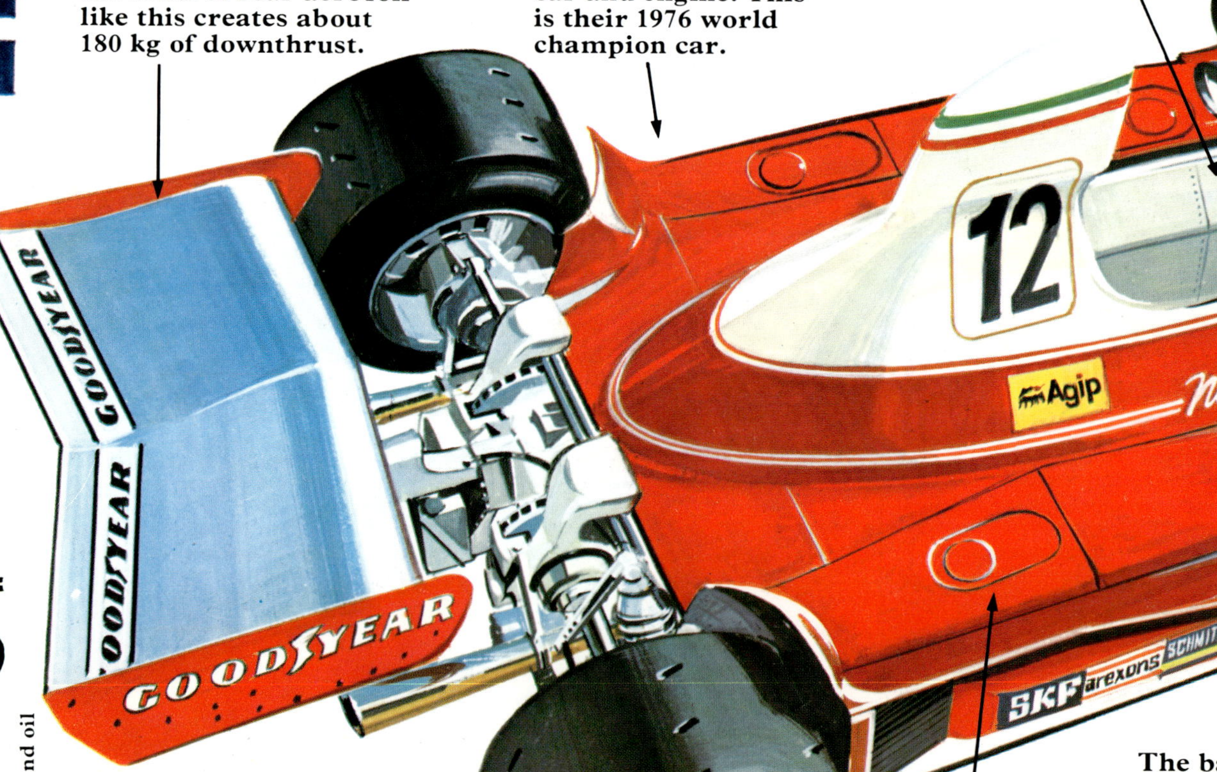

The back wheels are very wide, 510 mm. This helps them to transmit the power of the engine to the road with little wheel spin. The front wheels are narrower, 330 mm.

Fuel is contained in long tanks either side of the body

The basic frame of the car, made of steel tube and alloy sheet, weighs 35 kg. The complete car adds up to 585 kg.

Agip

Petrol and oil

HEUER

Precision timing equipment

Brakes

Shock absorbers

GOODYEAR

Tyres

The youngest driver to win the World Championship was Emerson Fittipaldi in 1972. He was then 25 years old.

Spark plugs

Petrol and oil

Shock absorbers

Sugar refiners

The engine burns a litre of petrol every 1½–2 km. The tanks hold 210 litres.

Front aerofoil

Tyres for wet and dry weather have different tread patterns and are made of different rubber mixtures. These smooth ones are dry weather tyres.

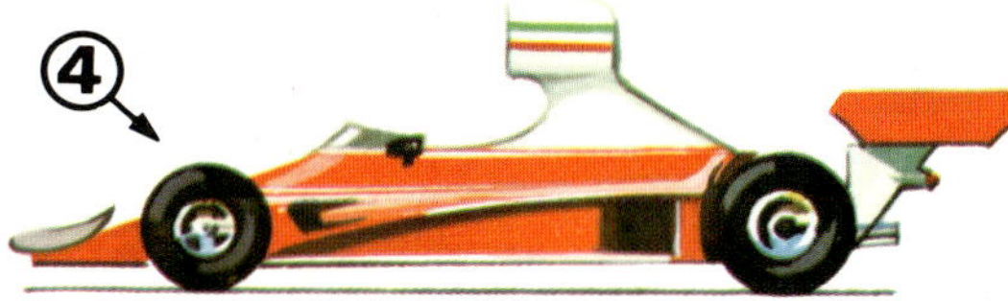

Aerofoils

All Formula 1 cars are equipped with aerofoils to prevent them lifting off the road at high speeds. Aerofoils are really like upside-down aircraft wings which, instead of lifting up, press down. This experiment shows the sort of effect that adding wings makes to a car's roadholding.

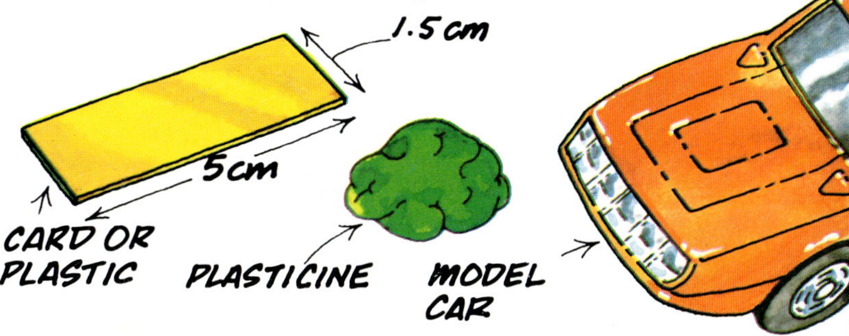

You need a model car, preferably made of metal. We used a 1/43 scale Renault Alpine. Cut out a rectangular aerofoil from card, or, better still from plastic. Try an old detergent bottle. You also need a lump of plasticine to fix the foil to the car's bonnet.

To equal the effects of high speed airflow on the race track you can use water, which is much denser than air. Fill up your bath, then send the car whizzing down the sloping end.

The car should sail down the bath smoothly. Dry the bonnet and add the aerofoil. Angled down, the car will stick to the bottom like glue. Angled up, the foil will lift the car (above)—a total disaster if it happened in real life. In fact, the first aerofoils were flimsy, dangerous affairs.

BAF

Airline

JCB

Earth-moving equipment

TISSOT

Watches

AP

Car spares

Petrol and oil

Ball bearings

Petrol and oil

Finance and banking

The brakes on the JPS 79 racing car can slow it down from 290 kph to 64 kph in just three seconds.

SPORTS RACERS

Fastest sports racer, the Porsche 917.

The world-beating sports cars of recent years have been made by the German firm of Porsche. The Porsche 935 shown below has been a winner at Le Mans and in the World Championship for Makes races. Le Mans is a 24 hour race in which the car which has covered the longest distance is the winner—in 1976 the 935 won the title, with new versions being developed the following year. For the North American CanAm races, Porsche developed the 917. It was the fastest road racer of all time, speed 413.6 kph.

The fuel tank holds 160 litres. Speeding on a twisty circuit at around 190 kph, the 935 uses a litre of petrol every 1.4 km.

This is the 935's fast circuit nose section.

The 935 has a top speed of 336 kph and can accelerate from 0 to 200 kph in just 7 seconds.

Different noses for different circuits

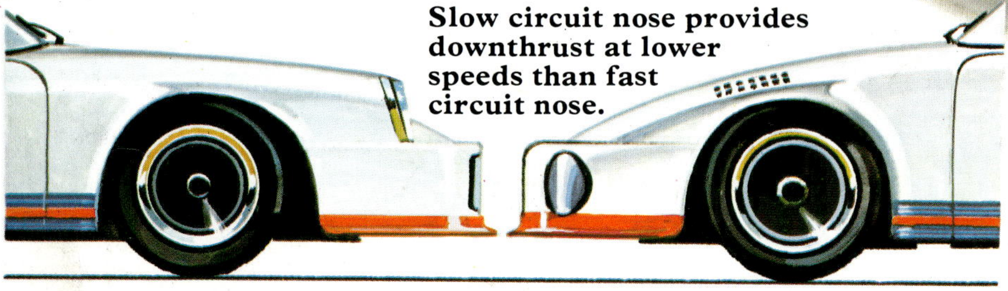

Slow circuit nose provides downthrust at lower speeds than fast circuit nose.

Porsche 935s have different nose shapes, depending on the track they race on. On slower tracks, the 'slow circuit' nose is used, shown above left. On fast tracks like Le Mans, drivers travel flat out and cars are equipped with lower, more streamlined nose sections. On circuits like this, the engines are higher geared too, allowing top speeds approaching 336 kph.

The Dunlop racing tyres need changing often—after only an hour on a demanding circuit. The disc brakes slow the car from 336 kph to under 80 kph in about five seconds. In that time, the car will have travelled 220 m.

The Chaparall 2J sports racer of 1970 was named the 'Vacuum cleaner'. It had a suction fan to glue it to the road round corners.

Porsche power

▲ Star test-car, the Porsche 911, powered by a 6-cylinder rear-mounted engine.

The first impression upon sitting in the driver's seat is that all the controls are in easy reach, with no awkward stretching or fumbling to reach odd switches. Electric-powered gadgets include windows, sunroof, radio aerial, and the angle of the door-mounted rear view mirror. This is controlled by a tiny 'joystick' on the doorsill and is a very useful safety feature.

Starting is instant, the roar of the engine booming out of the grille at the back. The engine warms up very quickly and there is little or no cold-engine jerk when moving off.

Around town, the 911SC is quiet and easy to drive—the stereo radio and sunroof make traffic jams a pleasure instead of an irritation.

Acceleration is fantastic—Porsche claim 0–100 kph in 7 seconds, though we only managed it in eight. The car corners as if on rails—the suspension keeps the car flat and stable but absorbs bumps well enough to provide a firm but smooth ride.

Claimed top speed is over 220 kph, and interestingly, the car gets quieter as it gets faster—the engine noise simply gets left behind, with little but a distant whistling noise to be heard.

There was one fault on our car—the speedometer was partly hidden by the steering wheel. All else was excellent, as it should be on a car costing around £14,000.

As petrol is used up, the balance of the car changes slightly. The driver has an adjustable suspension control in the cockpit to minimise the effects of this on handling.

Like Formula 1 racing cars the 935 has aerofoils to keep it firmly on the ground.

To meet the rules of the Le Mans and Championship for Makes races, the 935 must have the external shape of a production road car, in this case the Porsche 911. Within the basic shape, everything is designed for racing. The steel body is similar to the road car, but doors and various panels are made of weightsaving glass fibre.

AT THE RACE TRACK

If you go to a race meeting there are lots of things to spot.

Before the start, last-minute adjustments are made to the cars. This area is called the paddock and you almost always need a special pass to get into it. Once the race has started, repairs and refuelling are carried out in the pits, which are like small garages next to the track by the startline. In the 1908 French Grand Prix, real pits were dug in the ground so that mechanics could work under the cars. This no longer happens, but the name has stuck.

▲ **Major race timekeeping is carried out by electronic equipment. For individual cars and for practise, a three-chronometer system is used like the one above. The lap time is shown to the driver on a board.**

▲ **This board, held up in the pits to the driver as he flashes by shows: top row, his position in the race; middle row, how many seconds he is ahead of the eighth man; bottom row, how many seconds he is behind the sixth man.**

LAP CHART

Nº	Driver	Laps 1	2	3	4	5
11	J. SMITH	6	6	6	6	10
20	N. HAUTA	20	20	20	10	6
10	J. SHUNT	11	11	10	20	20
28	E. FILLIPE	10	10	28	28	11
6	M. ANDRE	28	28	11	11	28

▲ **Making a lap chart will help you know the positions of the drivers in a race. Fill in the drivers' names and numbers in the left-hand column. Every lap, note car numbers as they pass. In this chart, car 10 took the lead from car 6 on lap 5.**

▲ **There are two types of tyre, though there are lots of different makes. Dry weather tyres are smooth, with small nicks in the rubber so mechanics can see how they are wearing. Wet weather tyres have rain dispersing tread patterns.**

▲ **Watch out for the big vans carrying the racing cars. In this one, a car's disc brake system is being checked before its driver, Andrea de Cesaris, takes the wheel. The cars are light—a mechanic can lift the front end off the ground.**

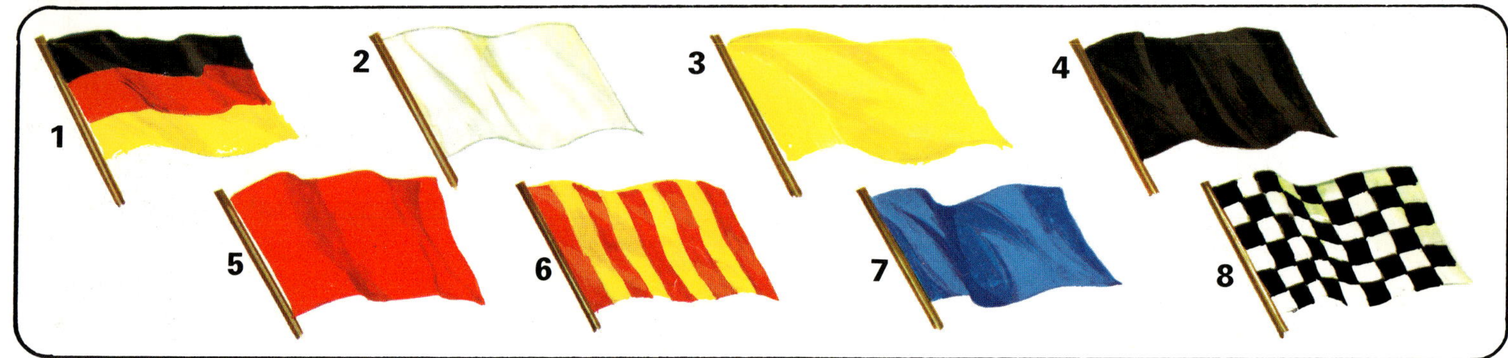

▲ **You will see various flags waved at a race meeting. Here are ones you are likely to see.**
1 Flag of the country where the race is taking place starts off the race.
2 Warning of ambulance or rescue vehicles on the track.
3 Held still, danger. Waved, more danger—drivers be prepared to stop.
4 Held with a board showing a car number. The driver must stop at the pits, perhaps to be told off for dangerous driving.
5 All cars must stop.
6 Slippery surface, probably oil.
7 Held still, another car close behind. Waved, the car is closing up fast or trying to overtake.
8 Waved for the winner. Held still to show other drivers that the race is over.

The fastest pit stop ever made was when Bobby Unser took only 4 seconds to take on fuel in the 1976 Indianapolis 500.

RALLIES

A rally is a form of time-keeping and endurance test. Cars leave the start at intervals, perhaps a minute apart. They then drive along a route, aiming to arrive at check points along the way at certain times. Teams lose points if they are late—and if they are early too. The Mercedes-Benz below won the London to Sydney rally of 1977.

► The six-and-a-half week race took the teams across Europe and the Middle East to India. The cars were shipped to Malaysia, then shipped again to Australia for the final leg of the rally. The red line shows the route they took, 30,000 km long.

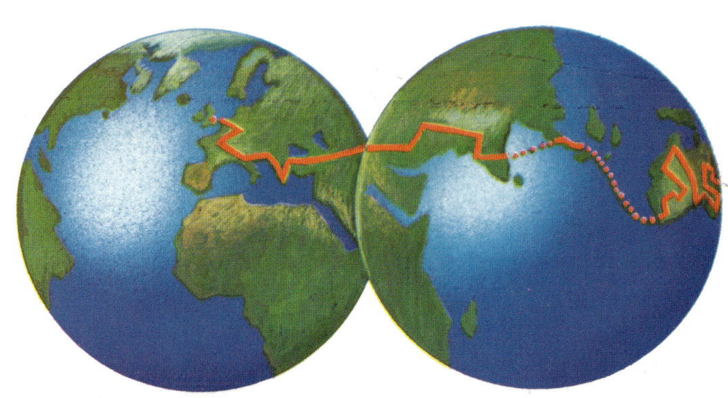

The inside of the car is equipped with a strong roll cage to protect driver and co-driver.

Emergency equipment includes drinking water, first aid kit, shovels, ropes and a spare radiator. The front bumper was useful—the car hit a kangaroo in Australia.

The suspension is strengthened to cope with the tough conditions.

Getting out of a sticky situation

The weather varied enormously during the course of the rally, from monsoons in India to desert conditions in Australia. The car was equipped with detachable bumpers, front and rear. If the car came off the road, they doubled as traction mats for the rear wheels to ride out of sandy or muddy ground. The picture above shows the rear bumper being taken off.

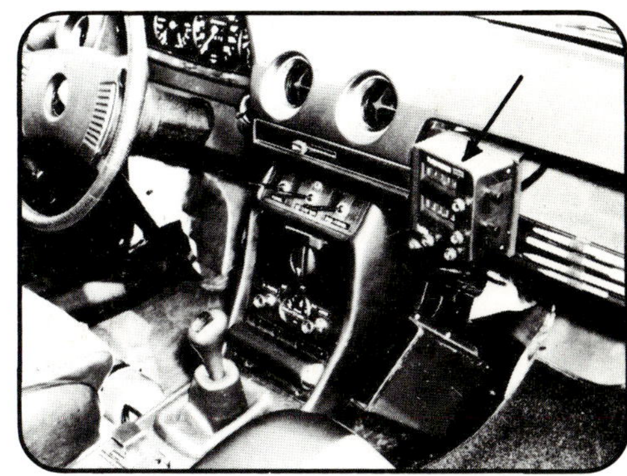

▲ This view of the car's interior shows an essential piece of equipment, the Halda Speedmaster, which calculates speeds and times along the route. Other equipment includes full seat belts, storage bins and pencil holders.

The Ford Escort is the most successful rally car, having won more than 76 international events in the last decade.

DRAGSTERS

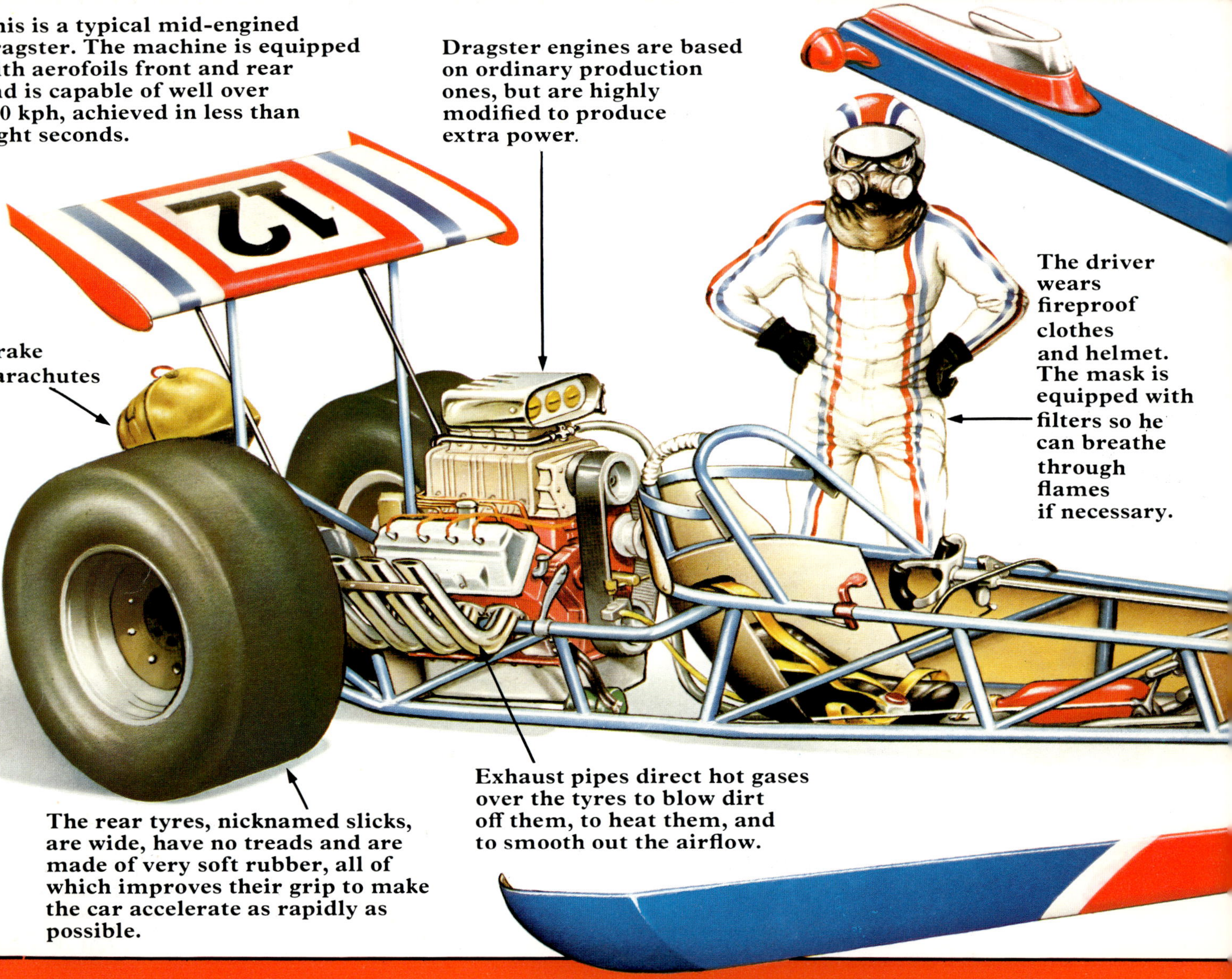

This is a typical mid-engined dragster. The machine is equipped with aerofoils front and rear and is capable of well over 300 kph, achieved in less than eight seconds.

Dragster engines are based on ordinary production ones, but are highly modified to produce extra power.

Brake parachutes

The driver wears fireproof clothes and helmet. The mask is equipped with filters so he can breathe through flames if necessary.

The rear tyres, nicknamed slicks, are wide, have no treads and are made of very soft rubber, all of which improves their grip to make the car accelerate as rapidly as possible.

Exhaust pipes direct hot gases over the tyres to blow dirt off them, to heat them, and to smooth out the airflow.

Dragsters race two at a time along a straight 400 m drag strip. The kings of the strip, the AA fuel dragsters, look more like skeletons than cars. Any excess weight (panelling around the engine for example) is left off, and they burn a special fuel called nitro-methane to get maximum power from the engine. A fuel dragster's engine develops nearly twice the power of a 'gasser', a gasoline (petrol) fueled car.

Although drag racing started in the USA, there are now strips in most countries across the world.

The 8-second, 400 metre dash

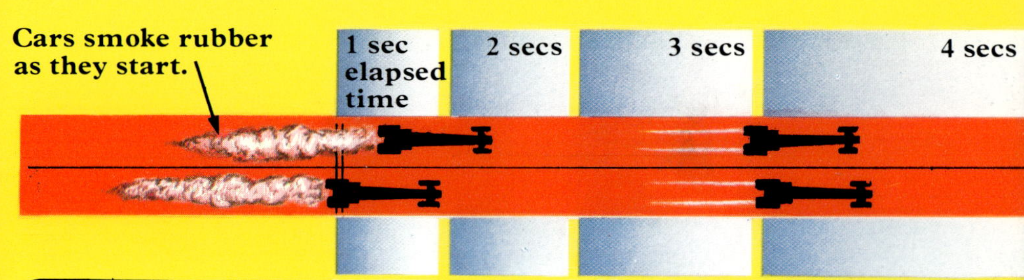

This will give you an idea of what happens in a typical drag race.

The competing cars are push-started in the 'fire-up' road. Then they make their burn-outs. For this, the drivers lock the front wheels and spin the back ones. The heat and friction of the spinning wheels warms the 'slicks', the back tyres, and lays down twin strips of tacky

The Ford Capri-based 'Eurosting' dragster uses nearly 23 litres of fuel for its 400 m run—a litre every 17 m.

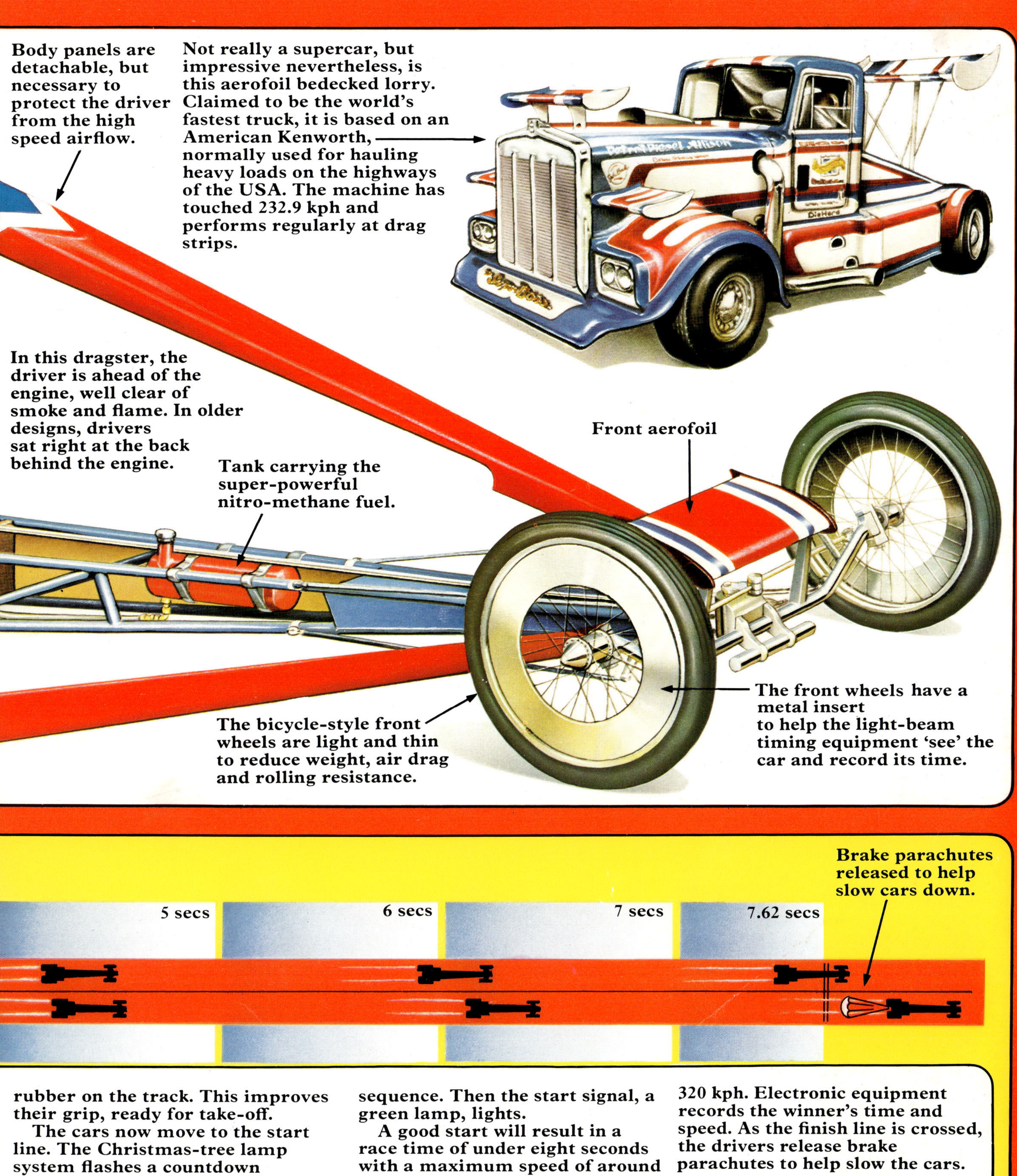

rubber on the track. This improves their grip, ready for take-off.

The cars now move to the start line. The Christmas-tree lamp system flashes a countdown sequence. Then the start signal, a green lamp, lights.

A good start will result in a race time of under eight seconds with a maximum speed of around 320 kph. Electronic equipment records the winner's time and speed. As the finish line is crossed, the drivers release brake parachutes to help slow the cars.

Rocket power is on the dragstrips. 'Blonde bombshell' uses an old missile motor to push it at up to 466.7 kph.

WEIRD WHEELS

On these pages you can see just six of the out-of-the-ordinary cars which have been made.

Other curious creations have included a motorized pram, made in 1922, equipped with a platform for the nursemaid to ride on at the back; the 1941 Le Dauphin tandem which could be petrol, electric or pedal powered and the Vultee aerocar of 1947, a small car with detachable wings to enable it to fly. There was also the Amphicar of the 1960s, capable of 108 kph on land and 11 kph in the water.

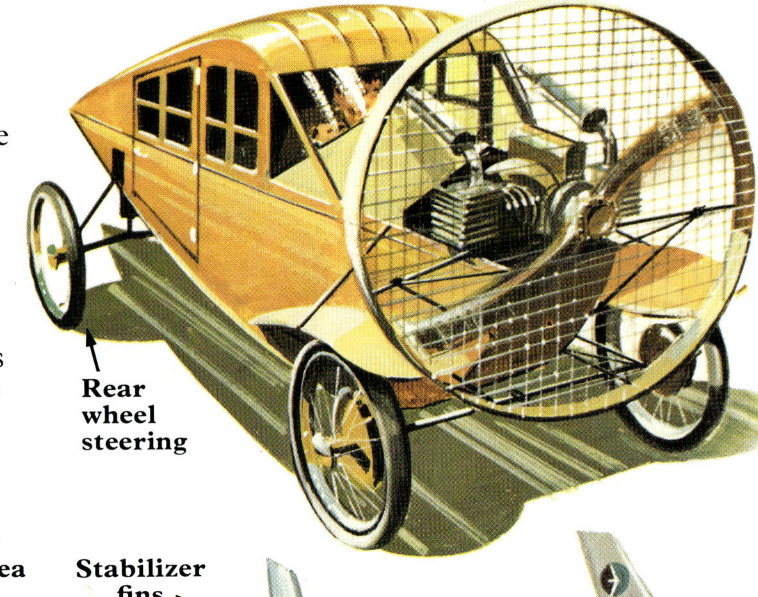

◄ This French Leyat Aerocar was built in 1923. It was pulled along at speeds up to 160 kph by its front mounted wooden propeller. Leyat must have been awkward to drive—it had back wheel steering and two brake pedals, one for each front wheel, so pressing them unevenly made the car swerve dangerously from side to side.

Single rear wheel **Rear mounted engine** **Passenger seating area**

▲ The vehicle above is the Dymaxion three-wheeler of 1933. It was designed by architect Buckminster Fuller. The overhang at the front and rear wheel steering made the car difficult to drive.

► These experimental Firebird cars were made by General Motors in the 1950s and 60s. They were powered by gas turbine aircraft engines and one of them had a single lever to replace the steering wheel, brake and accelerator pedals.

Fuel for a world on the move

All cars, even the odd ones on these pages, use fuel made from crude oil. Formed millions of years ago from the decayed remains of tiny plants and animals, most of it lies deep under the ground or sea bed. The picture sequence at the right shows the route oil takes, from oil rig to some of the end products. Each grade of oil has its own specialized uses.

The 1969 Lincoln Continental built for the American president had more than two tons of protective armour plate added to it.

▶ The Panther-6 was designed as a unique and expensive supercar. Its claimed top speed is over 320 kph. The prototype, shown here, is an open two-seater. Production cars will have powered folding tops to keep out the rain.

Aquaplaning and how to avoid it

Panther have found that the 6-wheel layout reduces the risk of aquaplaning. At low speed on a wet road, a tyre's tread pattern channels water out of the way, keeping the tyre in contact with the road. At high speed, too much water can build up to be channeled aside. Then the tyre rides up on a film of water. This is called aquaplaning and can result in total loss of control.

The Panther's front wheels clear water aside, so the second pair can get a good grip.

This diagram shows how the Panther's unique wheel layout allows safer driving in wet weather.

▲ This 'orange' car was built as a publicity stunt to advertise an orange importing firm. The shell was made of glass fibre, built up on the chassis of a Leyland Mini. Several have been made and the 'oranges' are fully roadworthy.

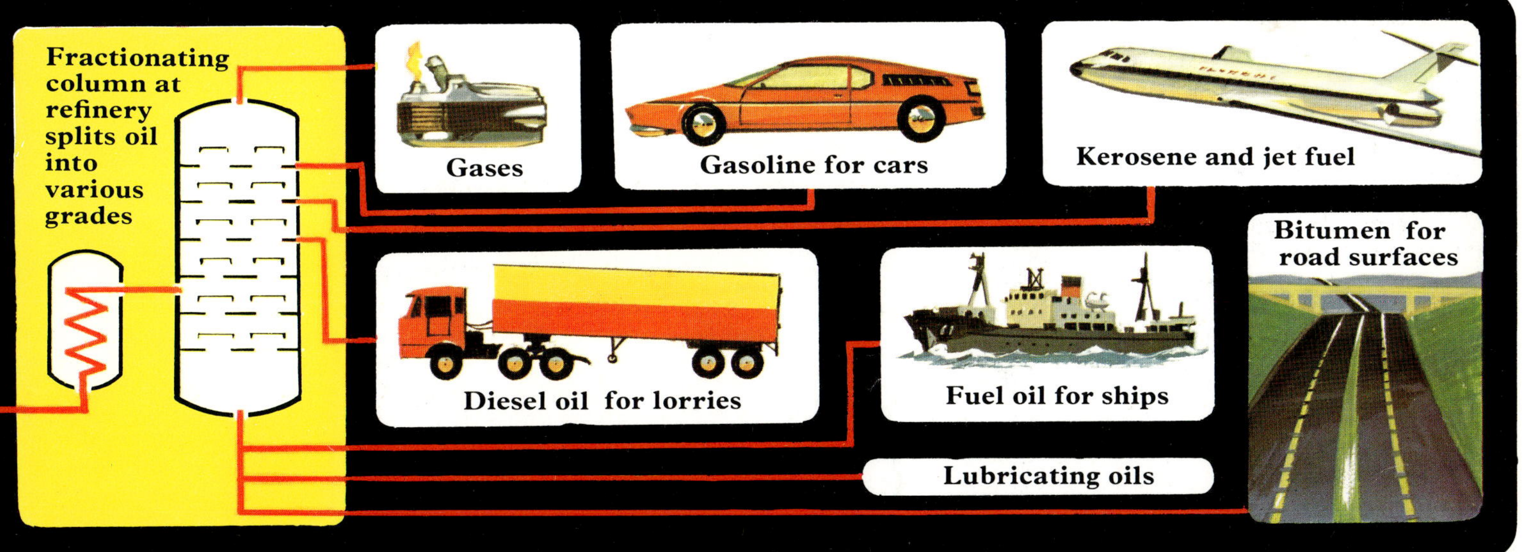

A tyre rolling on a wet road at 80 kph has to channel away over 5 litres of water a second to keep in contact with the road.

THE WORLD LAND SPEED RECORD

The tall tailfin is designed to keep Blue Flame steady at very high speeds.

Blue Flame was the first Land Speed Record holder to be powered by a rocket motor. The engine, mounted in the tail section, burnt a mixture of hydrogen peroxide and liquid natural gas. The car was the first to break the 1,000 kph barrier.

The long nose and rear cockpit layout of Blue Flame is typical of the latest generation of record breakers. The next goal is to go supersonic—over 1,200 kph.

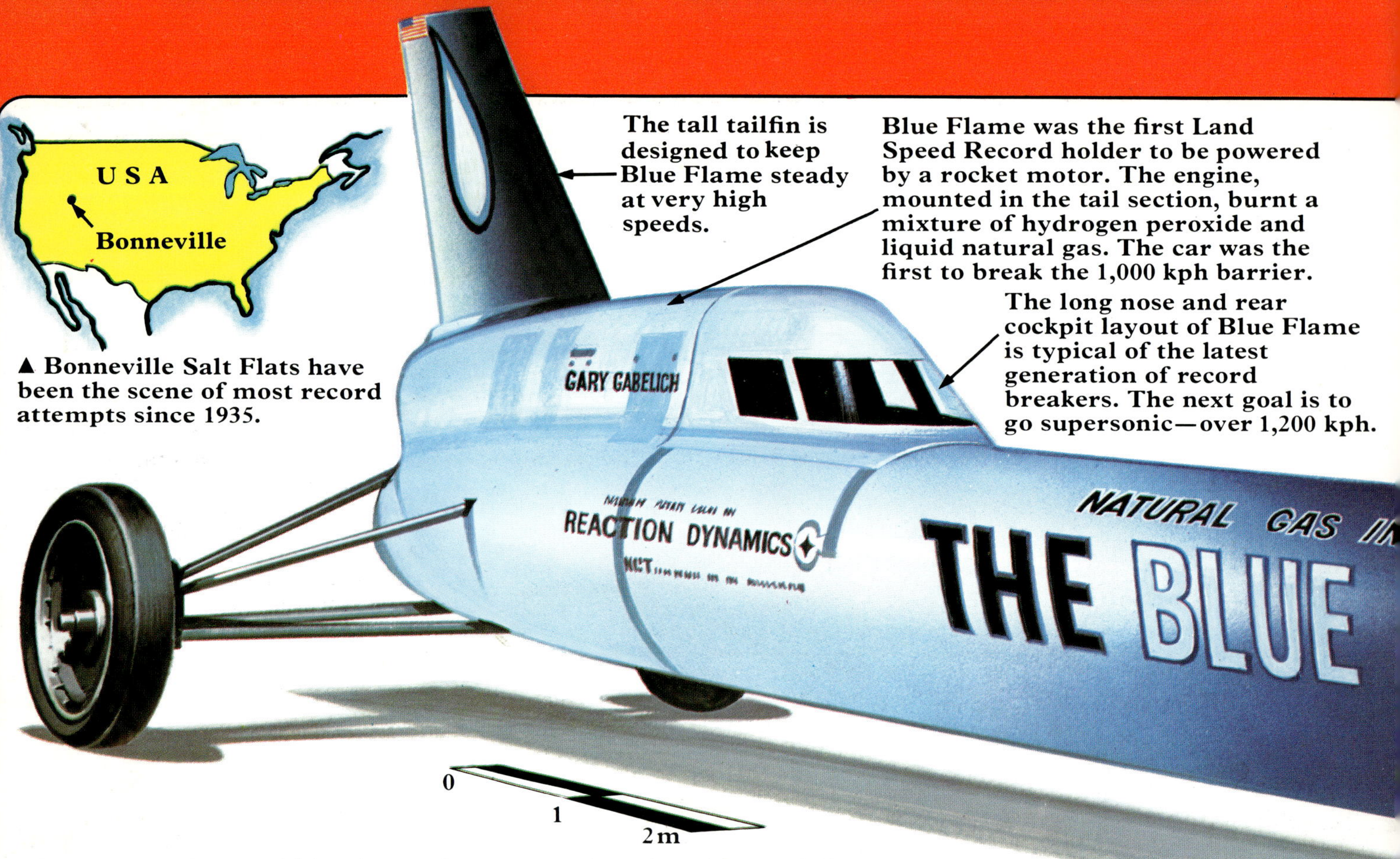

▲ Bonneville Salt Flats have been the scene of most record attempts since 1935.

Attempts at the world land speed record date from 18 December, 1898, when French driver Gaston de Chasseloup-Laubat drove an electric car in a park near Paris at the then fantastic speed of 63.15 kph.

At first drivers were thought to be in great danger if they drove at such speeds—being expected to die of heart failure or be unable to breathe.

Since then, speeds have risen to the current 1,014 kph, a record held by Gary Gabelich driving the Blue Flame rocket car shown on these pages.

Attempts are timed over a kilometre or a mile, with a flying start to build up speed.

Record breakers are now divided into several classes. They are: 1, vehicles with at least four wheels, driven through any pair of wheels. 2, vehicles with at least four wheels but not driven by them (jet or rocket powered). 3, vehicles like 2 but with less than four wheels.

Streamlining for speed

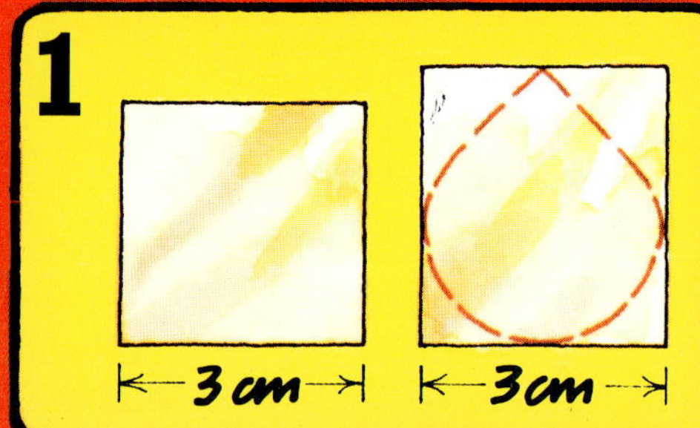

▲ Streamlining is the science of producing smooth sleek shapes to pass through the air with the minimum of resistance, or drag. This simple experiment uses water instead of air, but the principle is exactly the same.

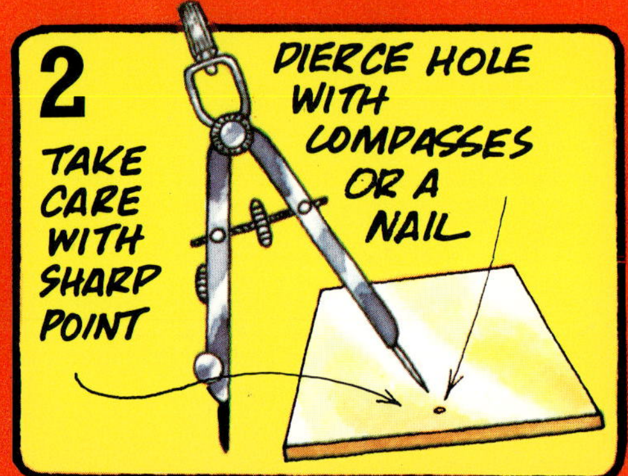

▲ Cut out two pieces of stiff cardboard to the shapes and sizes shown in frame 1. Then, using a pin or a pair of compasses, pierce a hole in the centre of each card about 1 cm from the front, as shown in the picture above.

▲ Cut two small cubes of plasticine about 1 cm square to act as weights. Each piece should weigh the same. Cut two pieces of thread 50 cm long and use these to attach a plasticine weight to each of the cardboard shapes, through the holes.

In 1966 Art Arfons crashed his 'Green Monster' car at 967 kph. It flew through the air, skidded and rolled, but he survived.

Blue Flame was 11.58 m long, making it the longest record breaker ever.

▲ In 1899 the bullet shaped electric powered Jamais Contente ('Never Satisfied') became the first car to exceed 100 kph. Beside it is Bluebird, the fastest wheel-driven car of all time. In 1964 it reached 716 kph.

Blue Flame is a four-wheeler, though the front pair of wheels are so close together they look like one fat one. The tyres were inflated to 24.6 kg per square cm—over 10 times as hard as those on an ordinary car.

Blue Flame may be the last record-breaker to have rubber tyres. New designs feature all-metal combined wheels and tyres as rubber ones tend to disintegrate at high speeds.

Shaped more like a high speed aircraft or space rocket than a car, Blue Flame's streamlined body allowed it to slip through the air with a minimum of drag.

▲ Fill up your kitchen sink with water, as high as it will go without overflowing. If you want to see the eddies made by the card shapes, you can sprinkle some tea leaves—or any fine powder that floats—evenly on the surface of the water.

▲ Carefully place both the square and tear-drop shape at the far end of the sink, with the plasticine weights dangling over the edge. It helps if you can get a friend to hold on to the weights while you position the card shapes.

▲ Let go! The streamlined shape will whizz past the square one, its smooth lines providing much less resistance to the water. You can compare the turbulence made by the square with the smoother ripples of the streamliner.

The White-Triplex record breaker of 1928 had no less than three aircraft engines, one in the bonnet, two at the back.

CAR DESIGN

The cars shown here are not intended as production machines, but are 'one-off' vehicles to display new engineering ideas and fashions in body styling.

Cars like these are often used as prestige vehicles at motor shows. Reactions from visitors are noted and popular features are sometimes put into production.

Probably the most famous car stylist was the Italian, Pinin Farina whose designs included many Ferraris. He died in 1966 but his design studio carried on in his name. The 'Modulo', shown on the right, is one of the team's most well-known design efforts.

This Pininfarina Modulo was produced in 1969. The car included many features unique at the time, such as the adjustable front aerofoil and the curious cut-outs for the upper sections of the wide, racing-style tyres.

Be your own car designer

The first rule when designing an exciting body shape is that there must be room inside for people. So asemble the manikin shown on the right as a first step. Trace off the body parts, transfer to a piece of stiff card. Cut out the shapes and assemble the model as shown in the circles. This manikin, together with the component shapes will help you design your car. On the page opposite you can see how these items have been used for a small sports design.

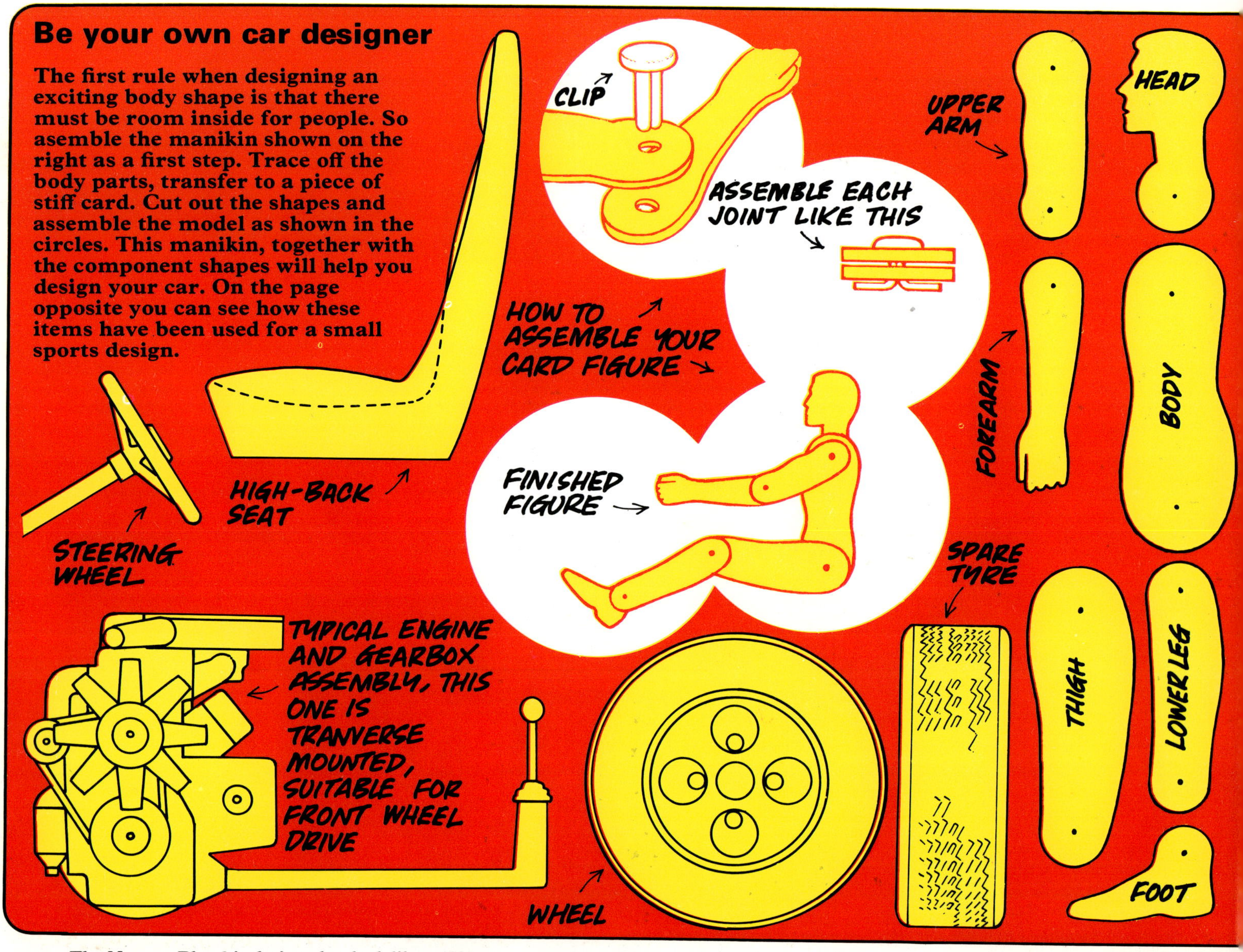

CLIP

ASSEMBLE EACH JOINT LIKE THIS

HOW TO ASSEMBLE YOUR CARD FIGURE

FINISHED FIGURE

HIGH-BACK SEAT

STEERING WHEEL

TYPICAL ENGINE AND GEARBOX ASSEMBLY, THIS ONE IS TRANVERSE MOUNTED, SUITABLE FOR FRONT WHEEL DRIVE

WHEEL

SPARE TYRE

UPPER ARM

HEAD

FOREARM

BODY

THIGH

LOWER LEG

FOOT

The Morgan Plus 8 is designed to look like a 1930s car, complete with a wood chassis. Its performance is modern—top speed 190 kph.

▲ These two designs were displayed at the 1978 Geneva Motor Show in Switzerland. At the top is an angular wedge-shaped sports model from the Japanese Dome design group. Below is the Megastar, based on the German Ford Taunus.

Three steps to a sports design

▶ The first step is to decide what type of car you want to design. For this one, we decided on a front-wheel drive, front-engine two-seater, with room for occasional passengers in the back. Good streamlining was essential, both to improve the car's top speed and its fuel economy. The basic components were sketched in and the driver model positioned.

▶ Here the design has progressed a little, with a basic body shape sketched in. The tiny back seats were abandoned in favour of a larger load space and the petrol tank was placed lower, protected by the crash bulkhead. The spare wheel was moved to the back which enabled the nose to be lower and sleeker. The manikin was placed in a more reclining position which enabled the roof to be lower and sleeker too.

▶ The finished design includes large lights front and rear and tinted glass to avoid glare from bright sun. The bumper strip all round gives valuable bodywork protection. The rear door frame hides a tough roll-over cage to protect the occupants in the event of a crash. Rear-viewing is taken care of by a wide-angle periscope built into the roof. The tyres have a reflective coating for high visibility from the side at night.

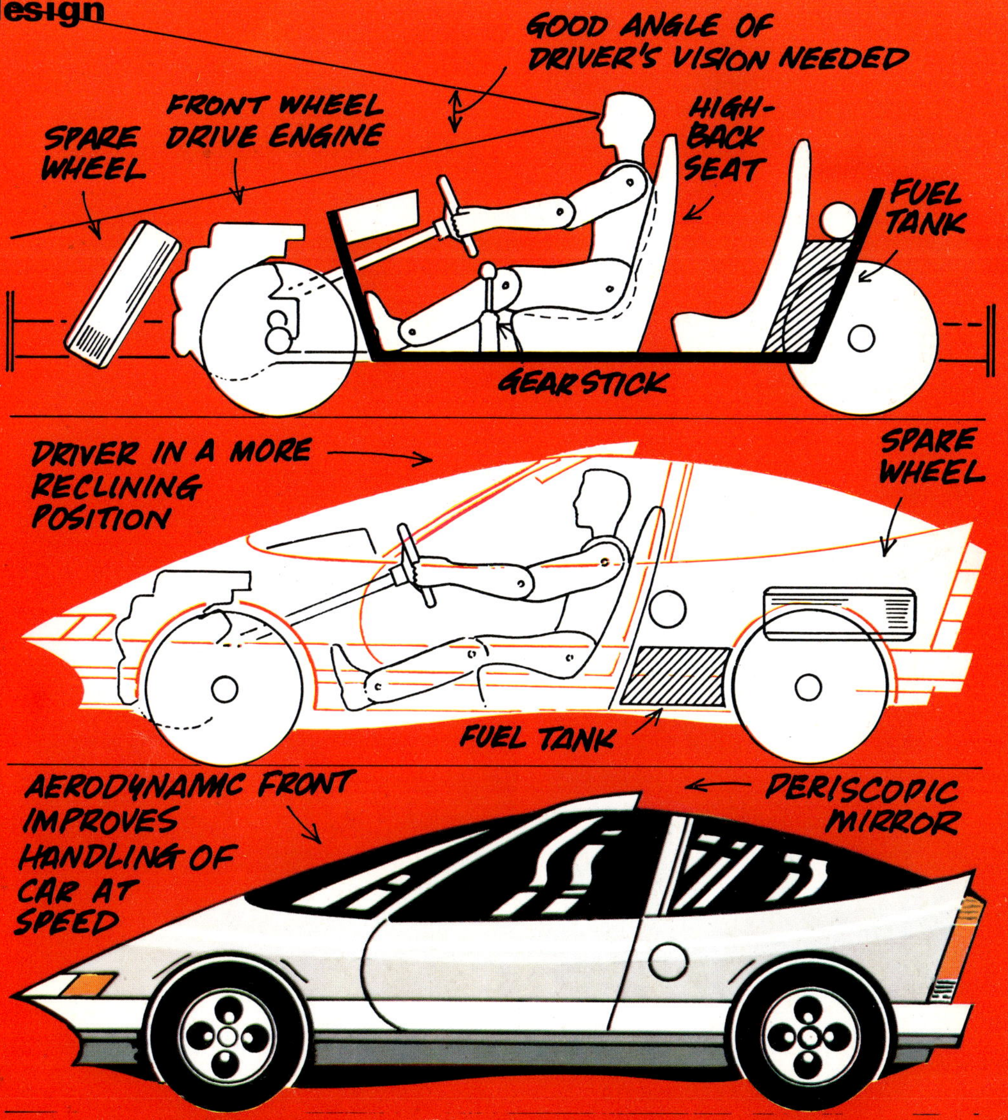

The largest engine ever built for a road-racing car was a 26.4 litre four-cylinder Dufaux of 1905.

MAKING A SUPERCAR

To find out how a supercar is built we visited the factory in Coventry, England, that produces the Jaguar XJ-S.

Many parts, like the electrical equipment and the body shell, are bought in from specialist manufacturers. The monocoque body is formed from huge sheets of steel in a giant press. 'Mono' means single, 'coque' means shell—the body and chassis are a single unit. The bodies are given a protective coat of grease before being sent to Jaguar. Most parts of the car are machine made, but assembled by hand on the XJ-S production line.

▲ **The bodies are dipped in degreasing and anti-rust baths. They then go to this primer paint shop. Each body is lowered onto a special frame which allows it to be tipped at any angle. It is then rubbed smooth between paint coats.**

▲ **This is the start of the production line. The paintwork is covered by protective sheets. About 150 men work on the 450 metre line, which moves forward at about 1 cm per second. In a 40 hour week the factory makes up to 100 XJ-S's.**

▲ **To fit the windscreen, the moulded rubber strip is first warmed to make it pliable. Once in place, the groove for the screen is painted with soft soap to make it slip in easily. The whole operation takes the experts two minutes.**

▲ **Each body is lifted onto a U-shaped overhead conveyor and taken into position over the raised second section of the line. Hanging over the line you can see the hoist control device used to lower the body gently on to the waiting axles.**

▲ **The engines are built in the nearby Daimler factory. Here, on the engine assembly line, the crankshaft is being fitted into a 12 cylinder engine block. The long bolts sticking out on the right are waiting to receive the cylinder head.**

▲ **This photograph was taken from below the line. The two 'baked-bean tins' are catalytic converters, fitted to cars going to the USA. They convert most of the poisonous exhaust fumes into less harmful substances.**

▲ **650 people make the upholstery for all Jaguar cars. They get through about 1,500 cow hides a week—2½ cows per car. To get the right shape, the worker first places a wooden template on the hide, and then cuts around it with a knife.**

▲ **To prepare for the final painting brown paper is used to mask windows and plastic guards are fitted over the bumpers and wheels. The car is wiped free of dust, blown with compressed air, then wiped again before the paint is sprayed on.**

There have been many car makers. In the USA alone, there have been more than 3,000 makes since the 1890s.

▲ The petrol tank is fitted between the boot and the rear seat—the safest place in a crash. Piping for the fuel supply has already been fitted. Many of the holes for pipes and fixing bolts are drilled as the car moves along the line.

▲ The large air conditioning unit has just been fitted in the middle of the fascia. It can keep the inside of the car at any temperature from 18° to 29°C. The complicated wiring system connected to the air conditioning goes in at this stage.

▲ Every engine is test run for one hour. Pipes and wires deliver water, oil, fuel and electricity, and remove exhaust fumes. The dials show the engine's performance. Up to 1,000 engines can be made and tested in a week.

▲ This picture shows the complete engine and gearbox being lowered into the body, which now has axles and wheels in place. It is such a tight fit that the engine must go in at an angle. One man works the hoist, two others guide the engine.

▲ The cars go from the paint shop into an oven for two hours. The temperature of 93°C hardens the paint. Each car was road tested before the final painting. After the oven, the cars go to the final inspection line, shown here.

▲ The XJ-S is complete. The 5,343cc engine gives a top speed of 246 kph, and acceleration from 0 to 160 kph in just under 17 seconds. The engine is so smooth and powerful that the car can pull away almost from a standstill in top gear.

Superselling cars

This chart shows the top six cars in the world league of supersellers. Sales of the Ford Model T averaged nearly 790,000 a year. The Beetle sold about 602,000 a year, but production continued much longer than for the Ford to put it at the top of the table.

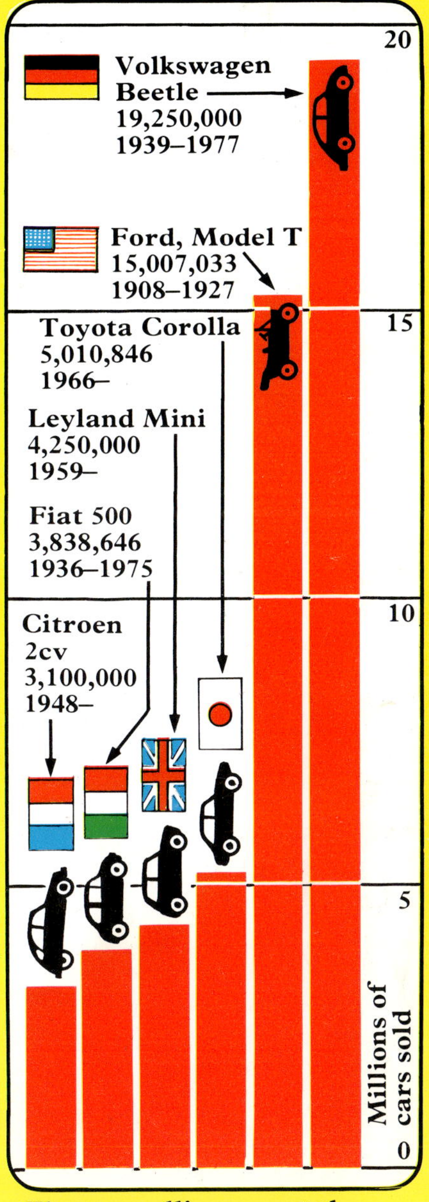

The top-selling car at the moment is the Japanese Toyota Corolla—in 1977, 729,901 were produced. Compare this with just 3,375 Jaguar XJS's made in the same year.

The world's widest car is Russian. The Zil 114 is 2.039 m across.

SUPERCAR 2000

The next twenty years are unlikely to bring any major changes in the design of the car. The story is probably going to be one of refinement, with special emphasis on safety, crash protection and fuel economy.

Electric cars will get more popular, but there will have to be a dramatic breakthrough in battery design to make them more efficient. Unless that happens, electric cars will not replace petrol-fuelled machines in large numbers.

These pages show you some of the ideas and research programmes currently underway.

Crashing to save lives

Laws in the USA state that the people in a car must be protected in the event of a head-on crash at up to 80 kph. Any cars sold there must meet these crash laws. Prototype cars are rammed into concrete blocks. This Rover 3500 has just passed an even tougher test—a head-on smash at nearly 100 kph. Yet the the passenger compartment remains intact .

Streamlining to save fuel

Petrol is going to get more and more expensive in the future, so designers wind-tunnel-test new designs to make them as streamlined as possible. The smoother a car slips through the air, the less fuel it needs to keep going. This design, by Vauxhall, is typical of the likely look for cars of the 1980s and '90s. Its features include an aerofoil (1), faired-in pop-up headlights (2), self-tinting glass (3), lightweight rustproof body (4), mid-mounted engine (5), four seats (6), lift-up tailgate (7), and a smooth underside to improve airflow under the car as well as over it.

Tyre makers Dunlop have devised a non-slip road surface called Delugrip. It could make skidding almost impossible.

Robots to save time

A computerized navigation system has been designed by the German firm of Bosch. It could make maps and map-reading a thing of the past. The destination is programmed into the machine via a control panel. The system picks up signals from wires buried in the road so it 'knows' where it is. The information panel (shown on the right) then tells the driver which way to go—straight on, turn left or right (central arrow) and the correct speed to drive at (number on the left). It also warns the driver of bad road conditions—ice, fog or heavy traffic.

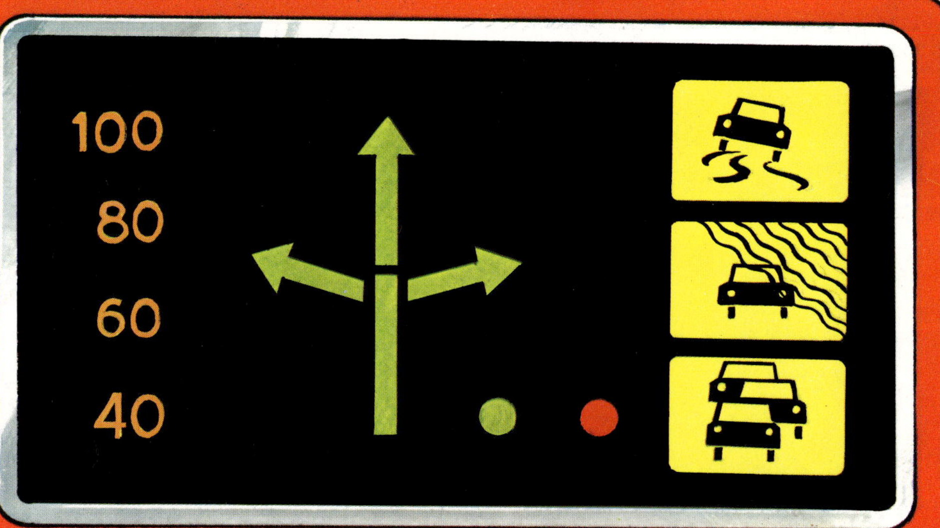

Engineering to improve design

This ESV—Experimental Safety Vehicle—shows the sort of skills which go into the design of a car. Based on the Chrysler Alpine saloon, it includes high mounted rear lights (1) to improve visibility. Specially strengthened doors (2) resist side impacts and headrests (3) protect peoples' heads and necks in an accident.

The windscreen (4) is shatterproof. A soft plastic nose (5) protects pedestrians and the bumpers (6) are designed to resist minor impacts without damage.

Electric runabouts to avoid pollution

As battery design improves, electric cars will get more popular. By the turn of the century, many (perhaps most) of today's mini-sized city cars could be electrically powered, being refuelled by plugging into the mains every night. Police cars like the one on the right could well have two engines—an electric one for cruising and a petrol one for emergency pursuits.

The Aston Martin Lagonda has a computer 'memory' for seat positions. Different drivers press a button and the seat adjusts itself.

RECORD BREAKERS

Over the years, vast amounts of talent and money have gone into raising the Land Speed Record to its present level of over 1,000 kph, yet curiously enough, there has been only one year in which a car has been the fastest vehicle of all.

In 1907 a steam powered car called the Stanley Wobblebug touched 241 kph before bouncing disastrously off the beach on which it was racing. Before then, trains held the speed record and ever since, aircraft and rockets.

Despite this, interest in racing and record breaking remains as high as ever.

Champion dragsters

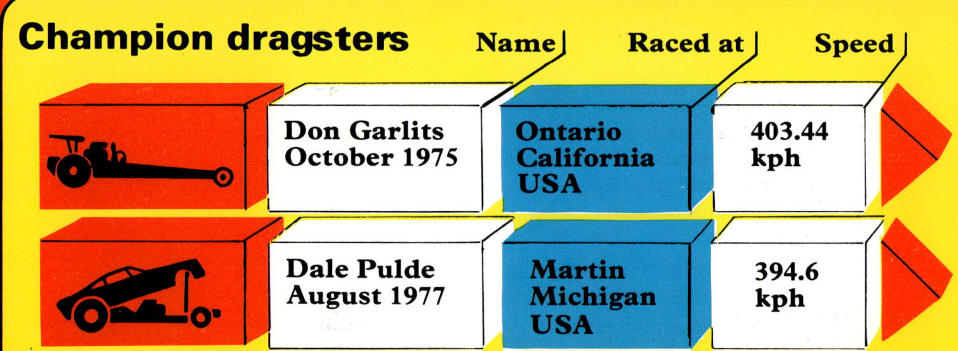

Shown here are the current top speeds on the drag strips. Don Garlits drove a Top Fuel dragster, one of the sport's 'Formula 1' cars, powered by a nitro-methane fuelled 9 litre engine.

Dale Pulde drove a Funny Car, one of the other major class of drag racers. Funny Cars have glass fibre bodies, to look like ordinary production cars, but they still look 'kinda funny'.

Champion tracks

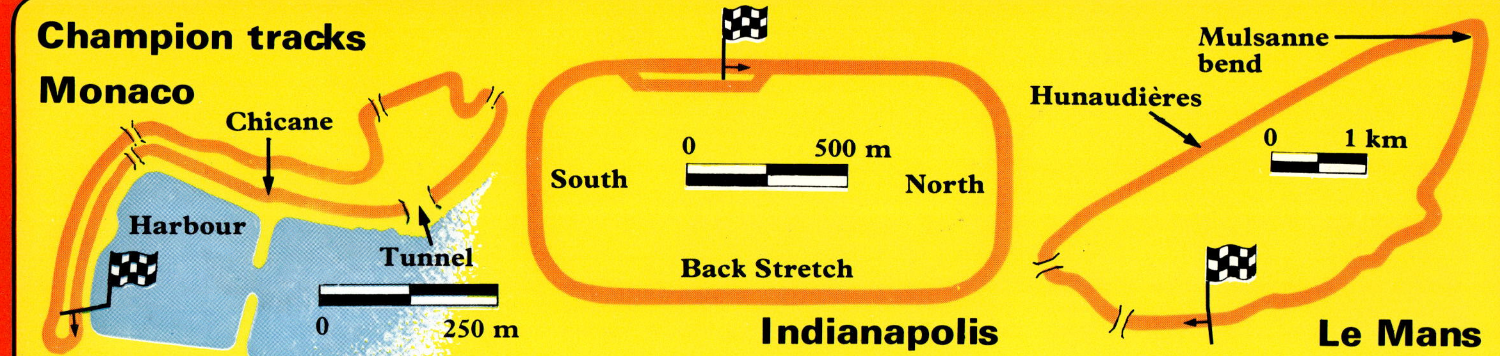

This track round the streets of Monte Carlo is the toughest GP circuit. It has 11 corners, a tunnel, several hills, a stretch of cobbles and no long straights. The record average speed for the race is 129.321 kph.

The best known GP race, the Indianapolis 500, takes place on this famous American circuit. In 1911, the winner averaged 120 kph over the 500 mile (805 km) race. Today the record is 262 kph.

This road circuit in France is the scene of the greatest endurance race, the 24 hour GP. The record distance covered in the 24 hours was achieved by a Porsche 917—a staggering 5,335 km.

Champion drivers

The chart on the right lists the world champions for every year since the Formula 1 championship began in 1950.

Although motor racing has been governed by restrictions since GP racing started in France in 1906, Formula 1 was only introduced after World War 2 when motor racing became very popular.

Racing formulas are revised from time to time. Since 1938 the they have been mainly concerned with engine size, and since 1961 safety has been the major concern.

Year	Driver	Car
1950	Nino Farina (Italy)	Alfa Romeo
1951	Juan Fangio (Argentina)	Alfa Romeo
1952	Alberto Ascari (Italy)	Ferrari
1953	Alberto Ascari (Italy)	Ferrari
1954	Juan Fangio (Argentina)	Maserati & Mercedes-Benz
1955	Juan Fangio (Argentina)	Mercedes-Benz
1956	Juan Fangio (Argentina)	Ferrari
1957	Juan Fangio (Argentina)	Maserati
1958	Mike Hawthorne (England)	Ferrari
1959	Jack Brabham (Australia)	Cooper-Climax
1960	Jack Brabham (Australia)	Cooper-Climax
1961	Phil Hill	

The Indianapolis 500 has big prizes. In 1974 $1,015,686 was handed out to winners of both the race and its various classes.

Champion speeds

The first land speed record breakers were driven by battery powered electric motors. Now record holders are all jet or rocket powered and expensive. Blue Flame cost no less than half a million dollars.
The chart below lists most of the important record breakers since the first attempt in 1898.

The fastest piston engined car

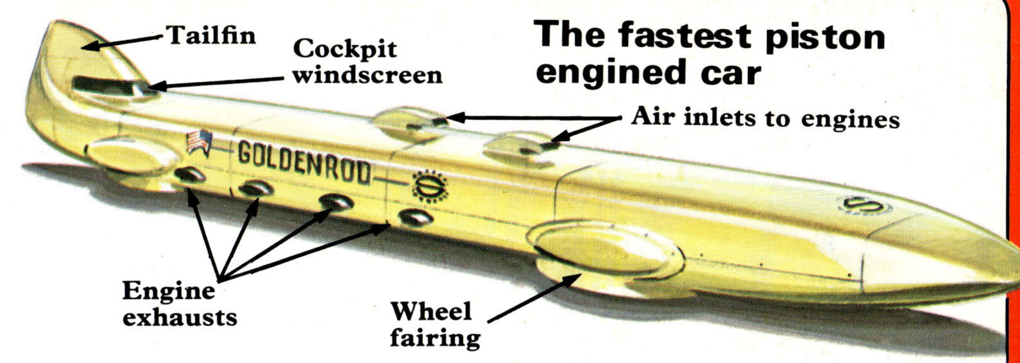

The picture above shows Goldenrod, which reached 673.5 kph on Bonneville Salt Flats in 1965, driven by Robert Summers. It looks little like an ordinary car, but its engines at least, were the same, though it had four of them, one to each wheel.

It was wheel-driven unlike record breakers like Blue Flame which are thrust along by rocket power. Donald Campbell's Bluebird was wheel-driven too, but its power came from an aircraft-type gas turbine engine.

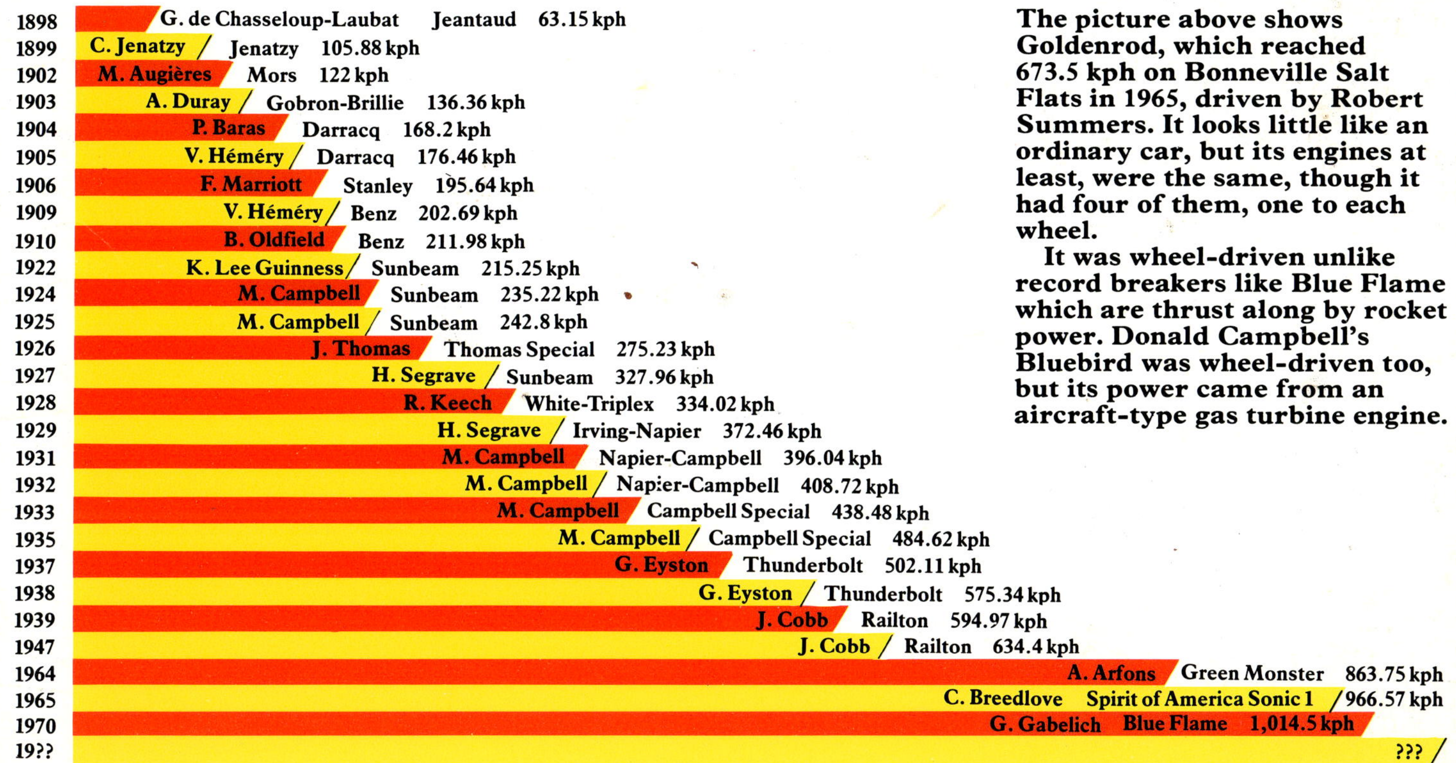

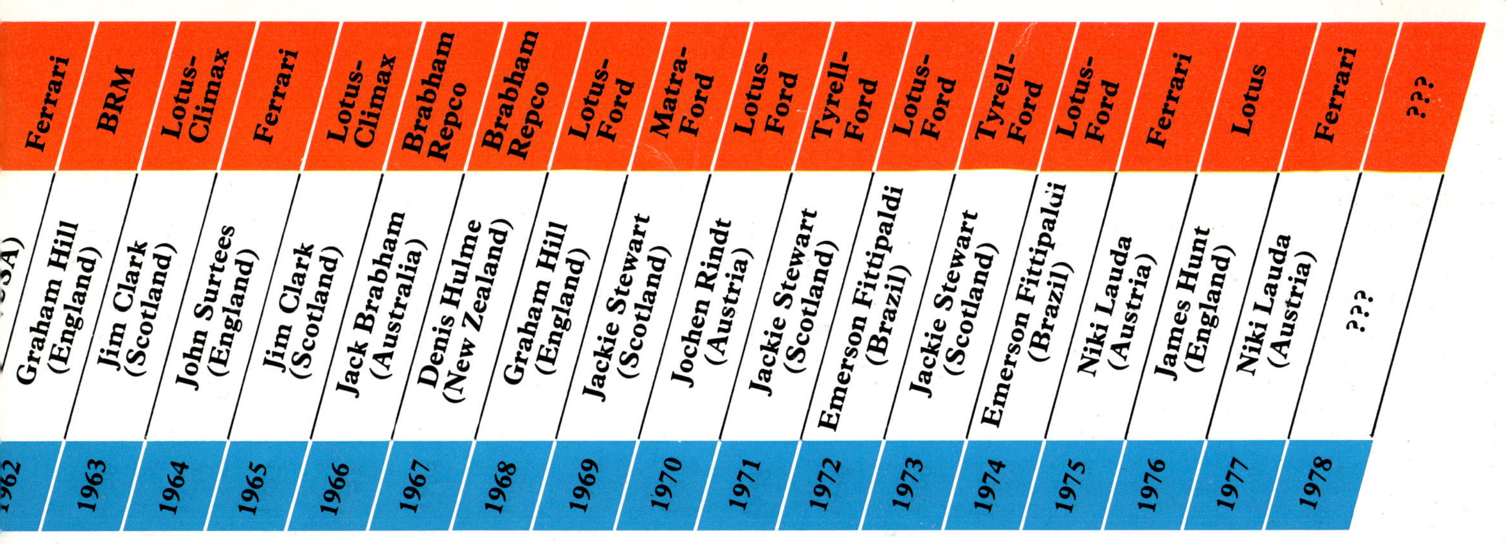

A car racing in the 'Indy 500' will use nearly 1,060 litres of fuel during the race.

INDEX

GOING FURTHER

Try building up your own motoring library. In addition to a good range of books, some of which are noted in the next column, you will need to keep up with the latest news and technical progress by reading weekly and monthly magazines. Here are some of the best available.

Autosprint, weekly, Italy. The superb pictures make this publication worth getting even if you do not read Italian.

Car, monthly, GB. Specially noted for its 'Scoop' feature—photographers take sneak pictures of the latest models while they are supposed to be secret, before manufacturers have announced them.

Car and Driver, monthly, USA. A wide-ranging and colourful magazine that is informative and attractively presented.

In addition to these glossy magazines, there are lots of weekly newspapers, and 'specials' on various motor racing events.

There are many books about cars. Some cover the history of motoring and some deal only with sports cars or racing cars. A few of the most useful books are listed below.

The All Colour Book of Racing Cars Brad King (Octopus)
Autocourse annual publication produced by Autocourse magazine (Hazleton Securities)
The Birth of the Motor Car Philip Sumner & Jenny Tyler (Usborne)
Early Motor Cars Anthony Bird (Allen & Unwin)
The Guinness Book of Car Facts and Feats Anthony Harding (Guinness Superlatives)
History of the Motor Car (New English Library)
Motor Car Christopher Tunney (Grisewood & Dempsey)
Sports Cars on Road and Track Ray Hutton (Hamlyn)
The World's Land Speed Record William Boddy (Motor Racing Publications)

Model making is a good way to have your own 'garage' of supercars. Here are some of the best kits, sold in most model shops.

Airfix, GB. Their range includes a 1/12 scale Bentley like the one shown on page 6. The model comes with green body parts, black rubber tyres and chrome coloured metal parts. It is big too—nearly 38 cm long when complete.

AMT, USA. Noted for their wide range of custom cars. Among the best is a 1/25 scale Chevrolet Corvette.

Revell, USA. A fairly small, but interesting selection includes a fascinating 1/4 scale model of a V-8 car engine. It has transparent sections so that you can see the various moving parts in action.

Tamiya, Japan. Make the world's best racing car kits, mostly to 1/12 scale. Super details like soft rubber tyres and a well-proportioned driver figure add to the realism of the kits.

If all the Beetles ever made were put end to end in a line, it would stretch for about 77,500 km.